Power Systems

More information about this series at http://www.springer.com/series/4622

Gang Lei · Jianguo Zhu · Youguang Guo

Multidisciplinary Design Optimization Methods for Electrical Machines and Drive Systems

Gang Lei
University of Technology Sydney
Sydney, NSW
Australia

Youguang Guo
University of Technology Sydney
Sydney, NSW
Australia

Jianguo Zhu
University of Technology Sydney
Sydney, NSW
Australia

ISSN 1612-1287 ISSN 1860-4676 (electronic)
Power Systems
ISBN 978-3-662-49269-7 ISBN 978-3-662-49271-0 (eBook)
DOI 10.1007/978-3-662-49271-0

Library of Congress Control Number: 2016930283

Printed on acid-free paper

This Springer imprint is published by SpringerNature
The registered company is Springer-Verlag GmbH Berlin Heidelberg

Preface

Electrical machines and drive systems account for about 46 % of all global electricity consumption, resulting in about 6,040 Mt of CO_2 emissions. This is by far the largest portion of electricity use, easily outstripping lighting, which takes up to 19 % of the world's demand. Therefore, the energy efficiency of electrical drive systems is very important for the energy conservation, environment and sustainable development of the world.

Electrical drive systems are key components in many modern appliances, as well as industry equipment and systems. In order to achieve the best design objectives, such as high performance and low cost, various optimization methods have been developed for design optimization of electrical machines and drive systems. The traditional design optimization is at the component level, e.g. optimization of a motor design or the parameters of a control algorithm. However, modern appliances or systems demand that the drive systems be specifically designed and optimized to provide full support to their best functionalities with multiple performance indicators. For such applications, the authors developed an application-oriented multi-objective system-level design optimization method. Because of the complexity of drive system design that involves many disciplines, such as electromagnetics, materials, mechanical dynamics including structural, thermal, and vibrational analyses, power electronic convertors, and control algorithms, a multi-level optimization method was developed by the authors to improve the effectiveness of the optimization of electrical machines as well as drive systems.

On the other hand, the real quality of motors and drives in mass production highly depends on the available machinery technology and those unavoidable variations or uncertainties in the manufacturing process, assembly process and operation environment. The manufacturing precision and tolerances are two main issues in the manufacturing process, including mainly the variations of material characteristics, such as magnetization faults in terms of magnitude and magnetization direction for permanent magnets (PMs), and density and permeability of soft-magnetic-composite (SMC) stator cores manufactured by powder metallic moulding technology, and dimensional variations of parts of drive systems, such as the rotor, stator, winding and PMs. The assembly process variations mainly include the lamination of silicon steel sheets and misalignments of stator, rotor and PMs. The operating uncertainties mainly include the load variations, changes of electrical and mechanical parameters, such as the changes of resistance and inductance due to the operational temperature rise, and fluctuations of drive voltage.

Limited by these variations in the practical machinery technology, an aggressively optimized design may be difficult for high-quality batch production and end up with high rejection rates. Similarly, variations in system parameters and operational conditions may also lead to sub-optimal performance, and in a severe case, even instability. To solve this type of problems, the methodology of Six-Sigma quality control can be adopted to develop a robust design optimization method to guarantee the high-quality batch production of drive systems.

Based on many years of research experience of the authors, this book aims to present efficient application-oriented, multi-disciplinary, multi-objective, and multi-level design optimization methods for advanced high-quality electrical drive systems. The multi-disciplinary analysis includes materials, electromagnetics, thermotics, mechanics, power electronics, applied mathematics, machinery technology, and quality control and management.

This book will benefit both researchers and engineers in the field of motor and drive design and manufacturing, thus enabling the effective development of the high-quality production of innovative, high-performance drive systems for challenging applications, such as green energy systems and electric vehicles.

This book consists of eight chapters, based on our several research projects, and covering the aspects of electrical machines, drive systems, high-quality mass production and application-oriented design optimization methods.

Like most books, this book starts with an introduction in Chap. 1 to provide an overview of application fields of electrical machines and drives as well as the state-of-art design optimization methods for electrical machines, drive systems and high-quality mass production.

Chapter 2 presents an overview of the design fundamentals of electrical machines and drive systems. Design analysis models in terms of different disciplines (domains) are investigated in this chapter, such as the analytical models or methods for electromagnetic and thermal analyses, magnetic circuit model for electromagnetic analysis, finite element model (FEM) for all electromagnetic, thermal and mechanical analyses, and field-oriented control and direct torque control algorithms for the control systems.

Chapter 3 reviews the popular optimization algorithms and approximate models used in the optimization of electrical machines as well as electromagnetic devices. Optimization algorithms include classical gradient-based algorithms and modern intelligent algorithms, such as genetic algorithms, differential evolution algorithm and multi-objective genetic algorithms. Approximate models include response surface model, radial basis function model and Kriging model.

Chapter 4 presents the design optimization methods for electrical machines in terms of different optimization situations, including low- and high-dimensional, single and multi-objectives and disciplines. Five new types of design optimization methods are presented to improve the optimization efficiency of electrical machines, particularly those PM-SMC motors of complex structures. They are the sequential optimization method (SOM), multi-objective SOM, multi-level optimization method, multi-level genetic algorithm and multi-disciplinary optimization method.

Chapter 5 develops the system-level design optimization methods for electrical drive systems, including single- and multi-level optimization methods. Not only the steady-state performance parameters but also the dynamic motor performance parameters, such as output power, efficiency and speed overshoot are investigated at the same time.

Chapter 6 presents a robust approach based on the technique of Design for Six-Sigma for the robust design optimization of high-performance and high-quality electrical machines and drive systems for mass production. A multi-level optimization framework is presented.

Chapter 7 develops the application-oriented design optimization methods for electrical machines under deterministic and robust design approaches, respectively. Applications including home appliance and hybrid electric vehicles are investigated.

Chapter 8 concludes the book and proposes the future works for further research and development.

Four electrical machines and several benchmark test functions/problems are employed throughout the book to verify the efficiency of those proposed design optimization methods. Those machines are a PM-SMC transverse flux machine, a PM-SMC claw pole motor, a surface-mounted PM synchronous machine and a flux-switching PM machine. All the design optimization models including FEM and thermal network model are validated by experimental results. Therefore, the proposed methods and obtained optimal solutions are reliable.

This book can be used as a reference for designers and engineers working in the electrical industry and undergraduate and graduate students majoring in electrical engineering. Students majoring in automotive engineering and mechanical engineering may also find this book useful when dealing with vehicle motor and drive related design, optimization and control development.

The authors wish to express their sincere thanks to Prof. Shuhong Wang, Xi'an Jiaotong University, China, for his contribution on the multi-level genetic algorithm for electrical machines and drive systems and other contributions to this book. The authors would also like to acknowledge the contributions of Dr. Yi Wang and Mr. Tianshi Wang on part of control algorithms, Dr. Wei Xu and Mr. Chengcheng Liu on part of PM flux-switching machines and multi-disciplinary design analysis of PM-SMC motors.

The authors would also like to thank their families who have given tremendous support all the time.

Finally, the authors are extremely grateful to Springer and the editorial staff for the opportunity to publish this book and help in all possible manners.

Contents

Abbreviations

AC	Alternating current
ANN	Artificial neural network
ANOVA	Analysis of variance
BLDC	Brushless direct current motor
CCD	Central composite design
CEMPE	Centre for Electrical Machines and Power Electronics
COP	Coarse optimization process
CSIRO	Commonwealth Scientific and Industrial Research Organization
CSRBF	Compactly supported radial basis function
DC	Direct current
DEA	Differential evolution algorithm
DFSS	Design for Six-Sigma
DOE	Design of experiments
DPMO	Defects per million opportunities
DTC	Direct torque control
EAs	Evolutionary algorithms
EDA	Estimation of distribution algorithms
EMF	Electromotive force
FEA	Finite element analysis
FEM	Finite element model
FOC	Field-oriented control
FOP	Fine optimization process
FSPMM	Flux-switching permanent magnet machine
GA	Genetic algorithm
GCD	Greatest common divisor
GEVIC	Green Energy and Vehicle Innovations Centre
HEVs	Hybrid electric vehicles
HTS	High-temperature superconductor
LCM	Least common multiple
LSA	Local sensitivity analysis
LSM	Least square method
MCA	Monte Carlo analysis
MDO	Multi-disciplinary design optimization

MLE	Maximum-likelihood estimation
MLGA	Multi-level genetic algorithm
MLSM	Moving least square method
MOGA	Multi-objective genetic algorithm
MPC	Model predictive control
MPSO	Multi-objective particle swarm optimization
MQ	Multi-quadrics
MSOM	Multi-objective sequential optimization method
NSGA	Non-dominated sorting genetic algorithm
NSW	New South Wales
PHEV	Plug-in hybrid electric vehicles
PM	Permanent magnet
PMSM	Permanent magnet synchronous machine
POF	Probability of failure
PSO	Particle swarm optimization
PWM	Pulse width modulations
RBF	Radial basis functions
RMSE	Root mean square error
RSM	Response surface model
SMC	Soft magnetic composite
SMES	Superconducting magnetic energy storage
SOM	Sequential optimization method
SPMSM	Surface-mounted permanent magnet synchronous machine
SVM	Space vector modulation
TFM	Transverse flux machine
UTS	University of Technology Sydney

Chapter 1
Introduction

Abstract This chapter presents a brief introduction focusing on various aspects of electrical machines, drive systems, their applications, energy usage, and the state-of-art design optimization methods. The design optimization of electrical machines and drive system is a multi-disciplinary, multi-objective, multi-level, high-dimensional, highly nonlinear and strongly coupled problem, which has long been a big challenge in both research and industry communities. The contents of this chapter form a good foundation for the whole book, and pave a smooth path to major goal of this book to present efficient design optimization methods for achieving high-performance high-quality electrical machines and drive systems for challenging applications, such as green energy systems and electric vehicles.

Keywords Electrical drive systems · Multi-disciplinary design optimization · Optimization methods · Electrical machines · High quality manufacturing · Mass production · Efficiency

1.1 Energy and Environment Challenges

In an electrical drive system, the role of electric motor is to convert electrical power into mechanical power. It is found that electric motors account for about 46.2 % of all global electricity consumption, leading to about 6,040 Megatons (Mt) of CO_2 emissions. This is by far the largest portion of electricity use, as shown in Fig. 1.1. Around the world, over 300 million motors are being used in industry, large buildings and infrastructure, and about 30 million new electric motors are sold each year for industrial purposes alone [1, 2].

Figure 1.2 illustrates the estimated electricity demand for all motors by sector. It can be seen that the motors used in industry consume about 63.1 % of the total energy consumption. The corresponding energy costs are estimated to be USD 362 billion per year. In the industrial sector, motors are used primarily for four areas of applications, namely pumps, fans, compressors, and mechanical movement. These applications and their respective shares are illustrated in Fig. 1.3 [1].

G. Lei et al., *Multidisciplinary Design Optimization Methods for Electrical Machines and Drive Systems*, Power Systems,
DOI 10.1007/978-3-662-49271-0_1

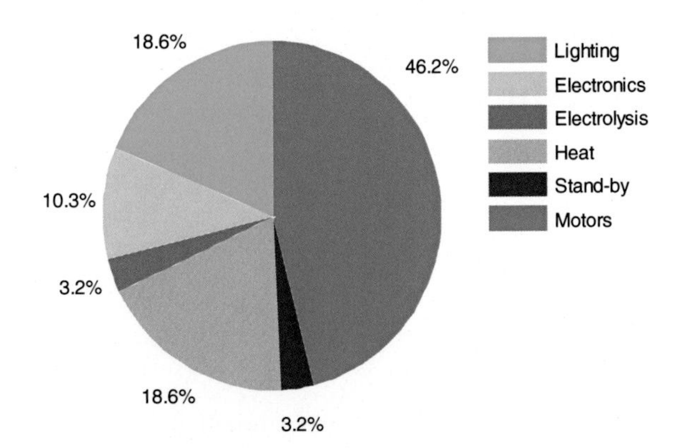

Fig. 1.1 Global electricity demand by sector and end-use

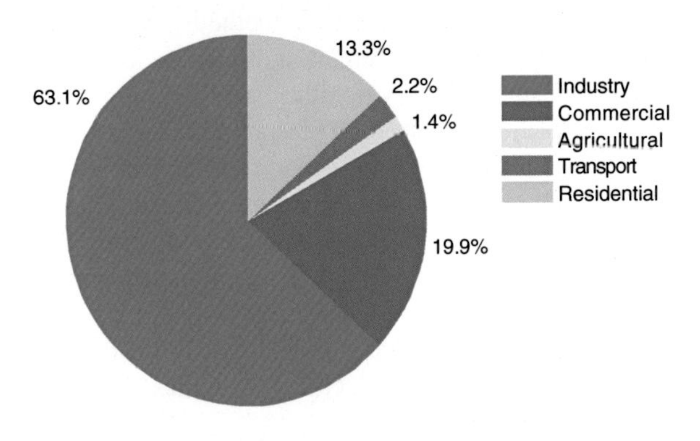

Fig. 1.2 Estimated electricity demand for all electric motors by sector

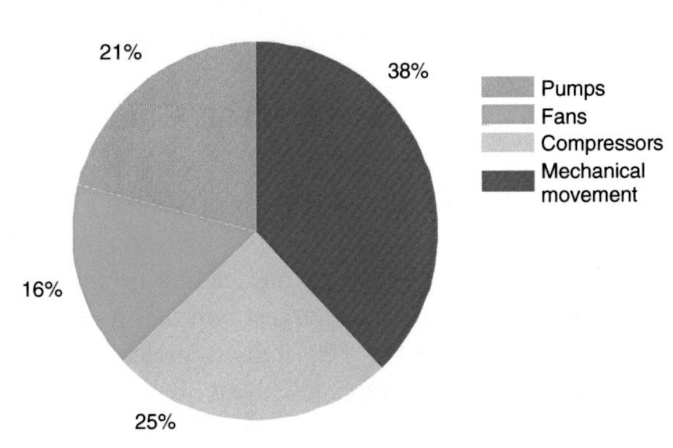

Fig. 1.3 Estimated demand by application in industrial motor system energy use

In terms of the life cycle cost of motors, the electric energy cost accounts for more than 90 % of all cost in general, which are much larger than the other two parts, purchase-price and repair or maintenance cost. As an example, Fig. 1.4 depicts the breakdown of the life cycle cost for an 11 kW motor operated 4000 h per year, where the electric energy cost accounts for 96.7 % of the total cost [2].

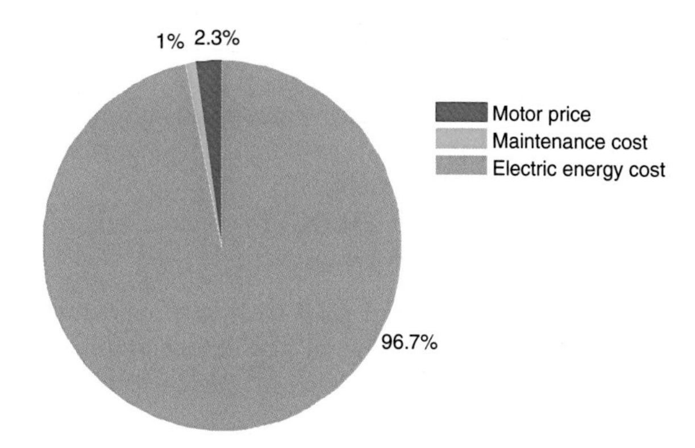

Fig. 1.4 Life cycle cost breakdown for an 11 kW motor operated 4000 h per year

Therefore, the energy efficiency of motors is a crucial issue for the energy conservation, environment, and sustainable development of the world, and this is also the main reason that high efficiency motors have attracted so much attention all over the world. Even 1 % increase in motor efficiency would save about 20 billion kWh per year or USD 1.4 billion in electricity and 3.5 million barrels of oil in the U.S. alone. These savings would be multiplied by about a factor of four on a worldwide basis [3, 4].

1.2 Introduction of Electrical Machines, Drive Systems, and Their Applications

1.2.1 General Classification of Electrical Machines

Electrical machines are electromagnetic devices for transforming electricity of one voltage or current to another voltage or current by the principle of electromagnetic induction for safe and convenient use (transformers), or electromechanical energy conversion (generators and motors).

Transformers can be generally classified as

- Power (step-up/down, and isolation) transformers, instrumentation transformers, and signal transformers by application,
- Three phase and single phase transformers by number of phases,
- Dry type or oil type transformers by cooling method, and
- Power frequency and high frequency transformers by application frequency.

Generators and motors can be classified as

- DC and AC machines by types of electricity they generate or are supplied,
- Rotating and linear electrical machines by motion style,

- Synchronous and asynchronous/induction machines by operational feature or principle,
- Round/cylindrical, salient, wound, and squirrel cage rotors by structure,
- Permanent magnet (PM) and high temperature superconductor (HTS) machines by construction materials,
- Brushless DC motor (BLDC) by combination of structural feature and operational characteristics, etc.

Since around 69 % of total electricity in industry is consumed by electrical motors worldwide, it is of great significance to use high efficiency motors and drive techniques. In a lot of applications, variable speed drive is more efficient than fixed speed drive.

1.2.2 Electrical Machines and Applications

In general, there are three main kinds of applications for electrical machines, which are electricity generation, electricity transformation, and electrical drives. The following are some examples.

A. Electricity generation

In most electricity generation systems, except the photovoltaic systems in which the solar energy is converted directly into DC currents by static solar cells, various types of energy resources, such as fossil fuels (coal, diesel, and natural gas, etc.), water, wind, sun light, and atomic energy, etc., are firstly converted into mechanical energy by rotating turbines and then electricity by AC rotating electrical generators.

Figure 1.5 shows the working principle of a hydroelectric generation with a synchronous generator. Hydroelectric plants use the energy from water to power a process that turns water potential energy into electricity. This process involves the water flowing from the dam, through a tunnel which leads to a turbine. Once the water reaches the turbine, the force from the water spins a generator to generate electricity. The generator terminal is connected to a transformer, which is where the electricity generated is transformed, e.g. to high voltage, and for long distance transmission [5].

Wind turbines as another application of AC generators have been employed worldwide. In general, both synchronous and induction machines are commonly employed for wind power generation. Moreover, various rotating AC generators are also commonly used for electricity generation from other energy sources, such as nuclear and solar thermal power plants, gas fired turbines, and diesel/petrol engines, to meet different needs, such as power supply for grid, air planes, trains, and ships.

B. Electricity transformation

Electricity transformers can be defined as a type of static electrical device which transfers power from one circuit to another by means of electromagnetic induction.

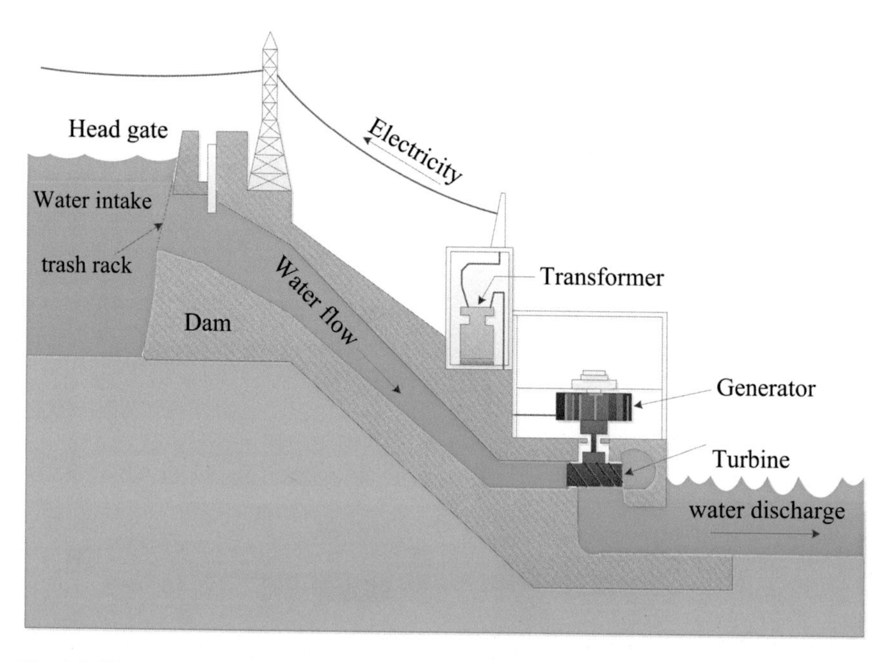

Fig. 1.5 Hydroelectric generation (synchronous generator)

Power transformers are used in distribution systems wherever there is a need to interface between different voltage levels, i.e. to step up and step down voltages. In order to improve the performance of transformers in renewable generation systems, some high-frequency transformers with cores of advanced magnetic materials, such as nanocrystalline and amorphous materials have been investigated recently. This offers a new route of step-up-transformer-less compact and lightweight direct grid integration of renewable generation systems [6–9].

C. Electrical drives

Electrical drives are a major application of electrical machines, which have been widely used in all aspects of our life. The following are some examples mainly investigated by the Green Energy and Vehicle Innovations Centre (GEVIC, formerly known as Centre for Electrical Machines and Power Electronics, or CEMPE) at the University of Technology Sydney (UTS).

Figure 1.6 shows a high efficiency (>90 %) 4 pole NdFeB PM brushless DC motor developed jointly by UTS CEMPE and the Commonwealth Scientific and Industrial Research Organization (CSIRO) for a submersible deep well (>120 m) mono-pump drive. The motor is filled with water and operated at 3000 rev/min. The product series has three power ratings of 300, 600, and 1200 W, respectively.

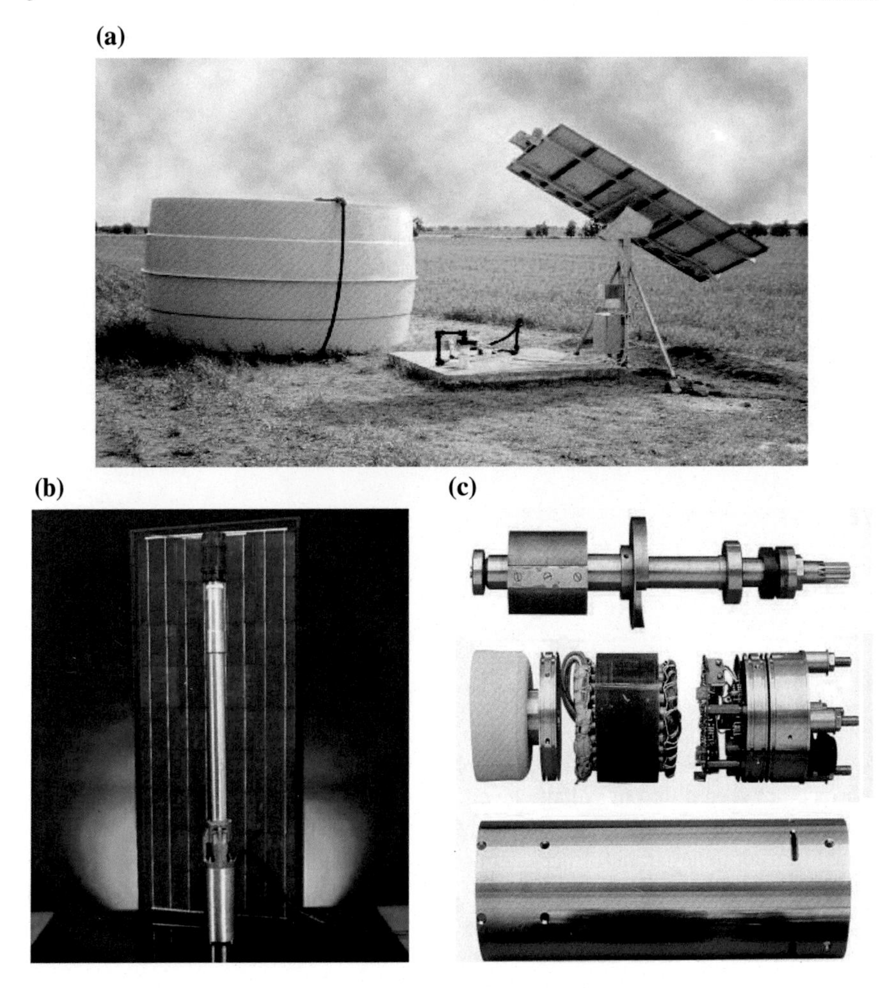

Fig. 1.6 A solar powered deep well submersible pump drive

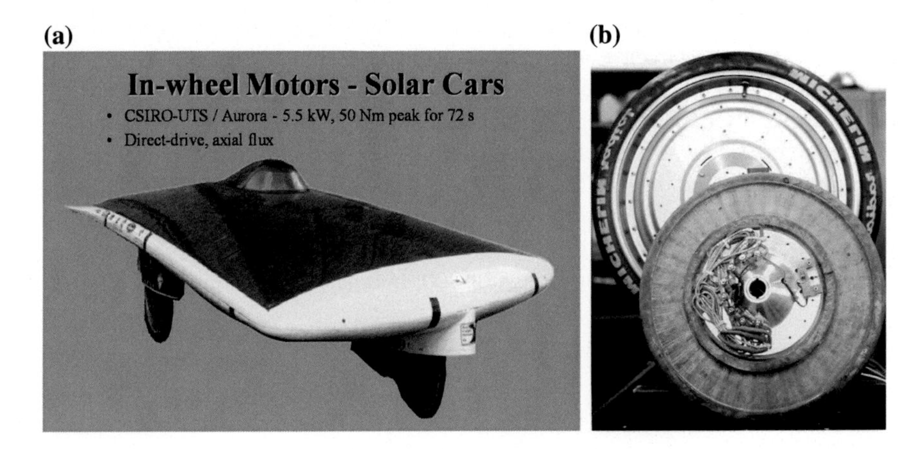

Fig. 1.7 **a** Aurora solar car and **b** its in-wheel permanent magnet motor

Fig. 1.8 UTS HEV test facility

Figure 1.7 shows a picture of the Aurora solar car and a photo of the in-wheel permanent magnet motor developed by UTS CEMPE and CSIRO. The ratings of the motor are 5.5 kW, 50 Nm peak for 72 s, with the maximum efficiency of 98.5 %.

In-wheel motors and other PM motors are widely used as drive machines in (plug-in) hybrid electric vehicles (HEVs). In a plug-in hybrid car, the battery bank is charged by the grid power supply when the car is not in use, and the electrical motor plays the major role of drive. A small internal combustion engine is employed to provide extra torque when the car accelerates, or to charge the battery when the state of charge is low. Since the motor controlled by a power electronic inverter can operate in all four quadrants of the torque-speed plane, the car is able to retrieve the kinetic energy by regenerative braking when it is decelerating. Therefore, hybrid electrical cars have much higher energy efficiency than the traditional internal combustion engine drive cars.

Currently, GEVIC researchers are designing several PM machines including flux-switching machine for plug-in HEVs. The designed and fabricated machines will be tested in the powertrain testing facility at UTS Automotive Laboratory, as depicted in Fig. 1.8. This facility can simulate urban and highway drive cycles and measure the torque and speed along with the powertrain performance in various operation modes, including regenerative braking [10].

Figure 1.9 shows the photos of the SolarSailor boat powered by 2 × 40 kW, 400 Nm, 3 phase, 16 pole, 950 rev/min, 100 V, 250 A direct drive high efficiency PM brushless DC motors and the power electronic controller developed by UTS CEMPE in 2000.

For propulsion and power supply of large ships, multi-MW electrical machines are used. In such cases, the efficiency, volume and weight of the electrical machines become a serious concern. With the technological breakthrough, large capacity HTS generators and motors are being built around the world for application in large cargo ships and gun boats.

(a)

(b) (c)

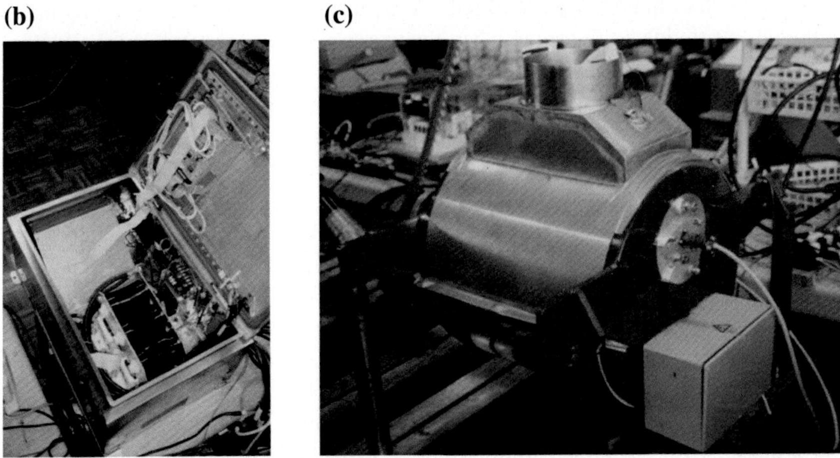

Fig. 1.9 **a** The SolarSailor boat, **b** power electronic converter, and **c** PM drive motor

1.3 The State-of-Art Design Optimization Methods for Electrical Machines and Drive Systems

1.3.1 Design Optimization of Electrical Machines

The design optimization of electrical machines is a multi-disciplinary, multi-objective, high-dimensional, highly nonlinear and strongly coupled problem, which is a challenge to both research and industry communities. This work consists of two parts: design analysis and performance optimization.

A. Design analysis

For design analysis, electromagnetic, thermal, mechanical design analyses and manufacturing design are the main concerns. The electromagnetic design is conducted by using mainly the analytical model, magnetic circuit model, and finite element model (FEM) to calculate the electromagnetic parameters of the machines being designed, such as flux linkage, back electromotive force (EMF), inductance and core losses. These parameters are then used to evaluate the machine performance indicators, such as output power and efficiency. The thermal design is conducted by using mainly the FEM and thermal network model to compute the temperature-rises in the machine. The mechanical design is often done by stress and/or modal analyses using the FEM to calculate the maximum stress, deformation, and resonant frequency of the machine structure under various operating conditions. In general, the thermal and mechanical analyses are usually conducted to check the insulation, magnetic strength (e.g. the Curie temperature of PMs), and mechanical material and structural strengths to ensure the machine's safety. In electrical machine design optimization, they are often used as design constraints.

B. Performance optimization

Performance optimization includes two aspects as well, namely optimization models and optimization methods. There are several popular optimization models. For example, from the perspective of objective numbers, optimization models can be classified as single-objective or multi-objective models. Generally, the cogging torque, torque ripples, material and manufacturing as well as sometimes operating costs, weight, and energy consumption are the main concerns in the design and optimization process. From the industrial perspective, the optimization models can consist of three main types: the deterministic, reliability and robust models.

Regarding the optimization methods, despite many kinds of optimization methods have been developed, their effective application for design optimization of electrical machines and systems has always been a research focus in electrical engineering. Since 1987, it has been selected as one of the most important development directions in computational electromagnetics by the premier International Conferences on Magnetics, such as Intermag (International Magnetics Conferences), CEFC (Conference on Electromagnetic Field Computation) and Compumag (Conference on the Computation of Electromagnetic Fields). In CEFC 2000, a special academic lecture about these problems was organized [11]. Electrical machines and drives are also an important section in several international conferences on electrical machines and energy systems, such as ICEMS (International Conference on Electrical Machines and Systems), ECCE (IEEE Energy Conversion Congress and Exposition) and IECON (Conference of the IEEE Industrial Electronics Society). Recently, the research interest in electromagnetic optimization design problems, particularly for electrical machines, has increased significantly. A special section on optimal design of electrical machines was presented in the IEEE Transactions on Energy Conversion in Sep. 2015 [12]. The main driving forces behind this interest are the rapid development of computational techniques and the rapid increase of industrial applications of electromagnetic devices.

Many electromagnetic optimization problems are solved by means of FEM with intelligent optimization algorithms. In the past two decades, a number of innovative intelligent algorithms, such as the genetic algorithm (GA), Tabu search, clonal selection algorithm, immune algorithm, particle swarm optimization (PSO) algorithm and differential evolution algorithm (DEA) have been developed [13–15].

The FEM can be accurate and applicable to nonlinear problems as well as general complex geometrical structures. However, it may not be appropriate to many design problems of electromagnetic systems because it is relatively complex and computationally intensive. As an alternative, some approximate models (also known as surrogate models) are employed in the practical engineering design problems to ease the computational burden of optimization process, such as response surface model (RSM), radial basis functions (RBF) model, Kriging model and artificial neural network model [13, 16–22].

C. Multi-objective optimization

From the perspective of practical engineering applications, the design optimization of electrical machines is actually a multi-objective problem as there are many objectives can be defined and one or some of them can be selected for different applications. For example, for home appliances, such as washing machines and refrigerators, the motor price and output power may be the two most important issues, while for hybrid electric vehicles, the volume, power density and torque ripples are very important. Therefore, multi-objective optimization design problems of electrical machines as well as other electromagnetic devices have become a topic of great interest recently. A few bench-mark problems have been proposed, such as TEAM benchmark problem 22 (superconducting magnetic energy storage: SMES) and Problem 25 (die-press model) [23–25].

In order to deal with these problems, many multi-objective optimization algorithms developed in the field of evolutionary computation have been employed, such as multi-objective genetic algorithm, non-dominated sorting genetic algorithm (NSGA) and NSGA II, and multi-objective particle swarm optimization (MPSO) algorithm. Meanwhile, some research works have been presented to improve these optimization algorithms, such as the improved NSGA, and improved MPSO [26–31]. A state of art multi-objective optimization methods in electromagnetism was presented recently in a monograph [32]. Approximate models have been employed to replace the FEM in multi-objective optimization problems to improve the optimization efficiency.

D. Challenges

For the above optimization methods, the direct optimization method based on FEM and intelligent algorithms are usually time-consuming and computationally expensive as a lot of FEM samples are needed in the optimization, especially for those machines with complex structures requiring 3D FEM and high-dimensional design parameters, such as the transverse flux machine (TFM) and claw pole motor. Moreover, premature is still a problem for all these algorithms though a lot of improvements have been made.

For example, the optimization process of a motor with 10 parameters (dimension D = 10) by using the GA and FEM with the population size of 50 (5 × D) and iteration number of 200 requires about 10,000 (50 × 200) samples, which can be a huge computational burden for many motors, especially those requiring 3D FEM.

On the other hand, it is impossible to replace the FEM with approximate models, such as RSM and Kriging model, because they cannot approximate high dimensional problems with sufficient accuracy by using reasonably small number of samples. For example, the first step in the construction of approximation models is to use the design of experiments (DOE) technique to obtain the initial samples. If 5 samples are required for each parameter, in total, 5^{10} FEM samples are required, which are more than those required by direct optimization method of GA&FEM [33].

Therefore, the traditional optimization methods based on FEM and the approximation models cannot solve the high dimensional design optimization problems. To solve this type of problems effectively, multi-level optimization methods developed in our previous work will be presented in this book [34–37]. The multi-disciplinary design optimization of electrical machines is still a challenge problem because of the huge computational cost of the coupled field analysis. A multi-disciplinary design optimization method will be presented for a PM TFM with soft magnetic composite (SMC) core in this book [38].

1.3.2 Design Optimization of Electrical Drive Systems

Electrical machines and the corresponding drive systems have a history of over a century and the design procedure has become almost standard. When designing an appliance that needs an electrical drive system, the designer firstly selects the motor, inverter/converter and controller from the existing products. The appliance designer, on one hand, has to deliver the functions that the appliance is supposed to have, and on the other hand, has to take into account the availability and performance that the existing motor drive can provide. This motor manufacturer-oriented approach has been the dominant design concept for drive systems for a long time. However, this approach would apply many constraints to the design and therefore limit the functions of the appliance [39].

With the fast development of numerical field analysis, CAD software, and flexible mechanical manufacturing technology, it is possible to design and manufacture a motor to meet the special requirements of a particular application such that the designer can concentrate on pursuing the best appliance functions. Since early 1990s, this application-oriented approach has been gradually becoming a common practice [40–48]. In many cases, the motor and control electronics are closely integrated into the appliances.

For example, the solar powered deep well submersible pump drive as shown in Fig. 1.6 that UTS CEMPE developed in 1991 has an integrated structure. The high efficiency motor and its electronic controller are packed into the pump and installed down the deep well. The concept of integrated design can be better illustrated by the in-wheel motor for the Aurora Solar Car drive (Fig. 1.7) developed jointly by UTS CEMPE and CSIRO in 1997. In this drive, to meet the special requirement of extremely low weight and high efficiency, a core-less in wheel motor topology was employed.

No doubt, the application-oriented integrated design concept is very advanced, but the design methodology used in these examples and also by all other motor designers is still very traditional. As illustrated in Fig. 1.10, the traditional design is conducted on the component level, i.e. the motor and the electronic controller are separately designed by the standard procedure. The major part of the motor design is the electromagnetic design whereas the thermal and sometimes mechanical analyses are carried out as verification only but not coupled to the electromagnetic analysis. In the case of design optimization, inaccurate circuit, field, and material models, and even empirical formulae are commonly used in order to reduce the computing time. This approach has two problems: (a) the real optimum design is not possible because of the inaccurate models used, and (b) the system performance cannot be optimized because the design is on the component level.

Meanwhile, the system performance optimization is becoming essential, especially when new materials, for example, the SMC, and new topologies are employed. SMC is a relatively new soft magnetic material developed for extremely low cost motor manufacturing using the highly matured powder metallurgical technology [49–53]. However, the magnetic properties of the SMC material are much poorer than those of the traditionally used silicon sheet steels. In order to develop low cost high performance SMC motor drive systems, we must explore new motor topologies of 3D magnetic flux and new drive schemes, and optimize the design at the system level.

The electric vehicles and HEVs are attracting great attentions and funding from the governments and general public around the world because of the worldwide fossil fuel energy crisis and severe greenhouse gas emissions of the conventional vehicles powered by the internal combustion engines. To improve the efficiency and drive performance with reduced volume, weight, and cost of novel drive systems to meet the challenging requirements of hybrid electric vehicles, a great amount of recent efforts are being directed towards the development and optimum design of high performance drive systems for (plug-in) HEVs [54, 55].

Through the extensive research practice, it is recognized that when designing such an electrical drive system, it is important to pursue the optimal system performance rather than the optimal components like motors, because assembling individually optimized components into a system would not necessarily guarantee an optimal system performance. The optimal system performance can only be achieved through a holistic approach of integrated simultaneous optimization of all components at the system level [39].

Figure 1.11 shows a brief design framework and the coupled relations of an electrical drive system. As shown, design optimization of such a system is a multi-disciplinary, multi-level, multi-objective, and high dimensional problem. It mainly includes electromagnetic, material, mechanical, thermal, and power electronic designs, which are strongly coupled [38, 39, 56, 57].

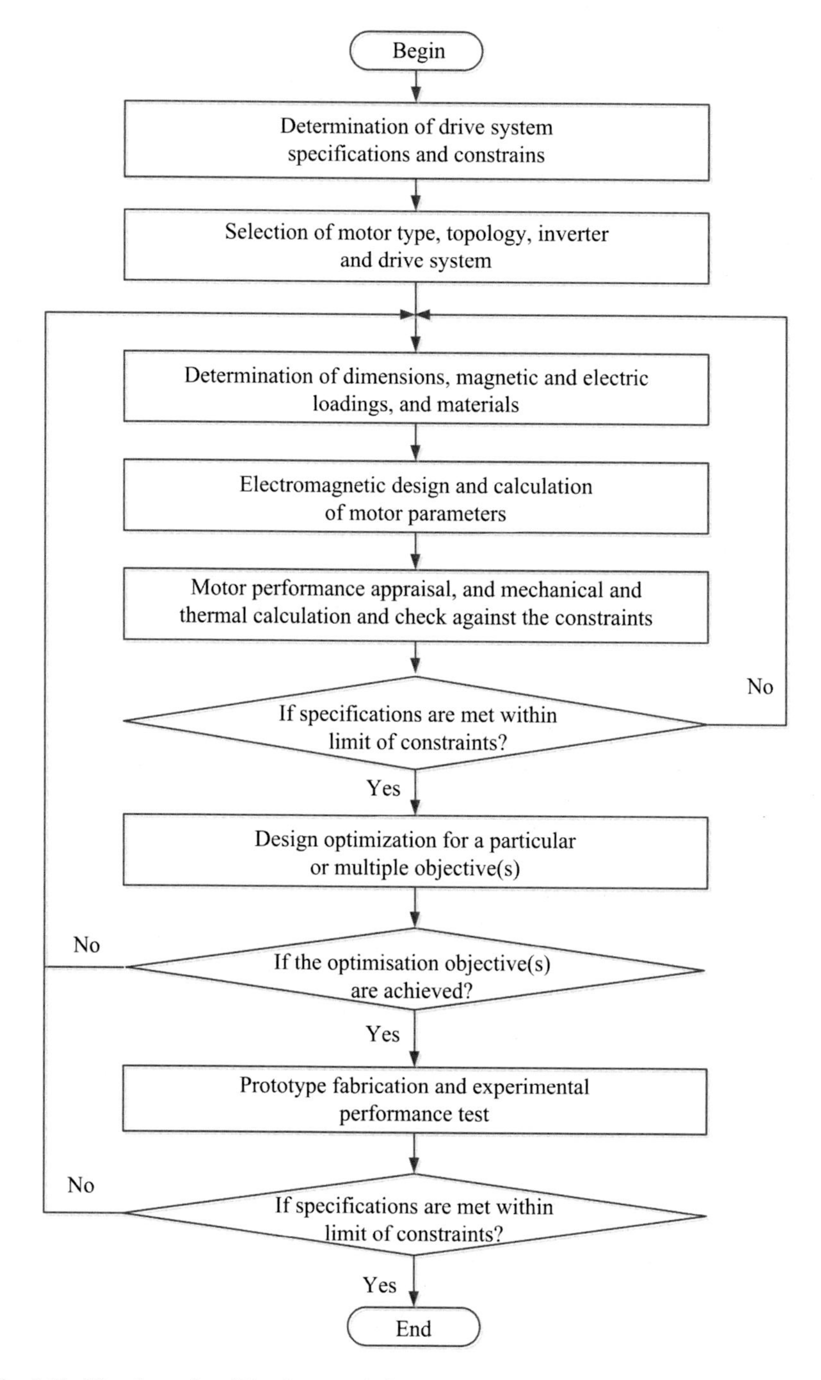

Fig. 1.10 Flowchart of traditional motor design procedure

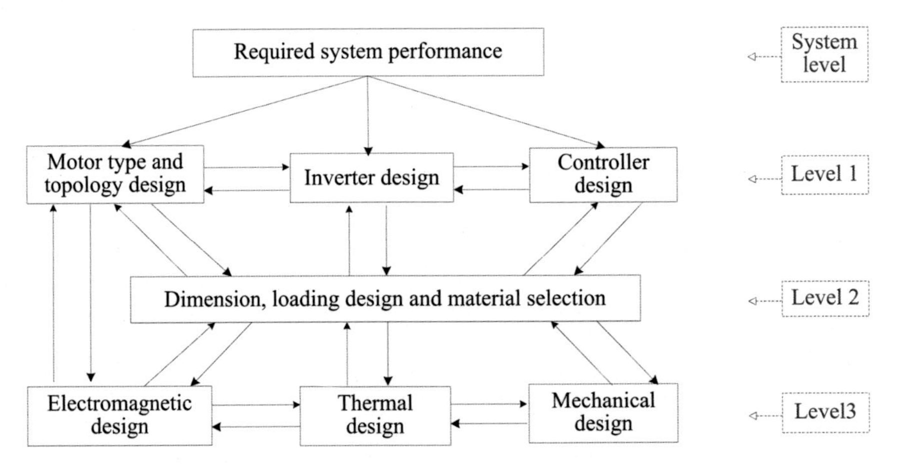

Fig. 1.11 Multi-disciplinary and multi-level design framework of electrical drive systems

Although the importance of system-level design optimization of electrical drive systems is well known, not much work has been reported in the literature [38, 39]. The traditional design and optimization methods are mostly on the component level of different kinds of motors [13, 16, 17, 34, 36–38].

On the other hand, regarding the controller design, though various control algorithms have been developed, such as the field oriented control (FOC), direct torque control (DTC), and model predictive control (MPC) [58–64], the design optimization remains on the component (controller) level, and is not combined with the motor design optimization [65].

This component-level-based method may be reasonable for some conventional motors and drive systems since there are a great amount of design experience and experimental test results. When designing a novel drive system, however, the designer does not have much design experience, and a holistic system-level approach becomes essential [39].

In this book, several types of system-level design optimization methods developed in our previous work will be presented and discussed for electrical drive systems.

1.3.3 Design Optimization for High Quality Mass Production

The design optimization method mentioned above are all deterministic design optimization approaches which do not take into account the unavoidable variations (noise factors) in the engineering manufacturing, including mainly the material diversity, manufacturing error and assembly inaccuracy, and system parameter variations in practical operation environment [66–68]. The motor and drive system performance depends highly on the manufacturing quality, which is in turn determined by the machinery technology or the manufacturing method and conditions. The variations or noise factors in the manufacturing process will result in big variation in motor and drive system performance. Limited by the manufacturing technology, an aggressively optimized deterministic design may be very difficult to make, and result in high rejection rates in mass production. Some details for the variations in manufacturing and assembly processes are discussed as follows.

A. Manufacturing process and tolerances

The manufacturing process and tolerances mainly include the material characteristics and dimensions of all parts of a drive system, such as the rotor, stator, winding, PMs and insulating material.

The manufacturing quality of PMs, for example, is crucial to the performance of PM motors. There are at least two kinds of variations in the manufacturing of PMs [66]. The first one is the dimension, such as the height and width, and the second one is the magnetization error of magnitude and direction. In [66], a practical example about the measurement data of PM width for a batch of 2,000 PMs was presented. These PMs were from three manufacturing groups with the same lower limit (14.60 mm) and upper limit (14.70 mm). The measurement revealed that the average of one group (1,000 PMs) is obviously smaller than the lower limit, and there is about 0.05 mm deviation from the average.

Figure 1.12 shows the manufacturing process of a stator core (one stack) designed for a claw pole motor made of the SMC material. Compared with the traditional silicon steel sheets, the motor cores made of SMC material are isotropic

Fig. 1.12 Low cost mold, press, and molded core before and after curing

both mechanically and magnetically, so that they are natural choices for the design of motors requiring 3D magnetic flux paths. Unlike the laminated cores made of the traditional silicon steel sheets, SMC cores can be manufactured by compacting SMC powders in a mold, and thus suitable for constructing motors of complex structures.

Figure 1.12 also shows a photo of the mold, press in the lab, and molded core of a three phase PM claw pole motor before (*white*) and after (*black*) thermal curing. On the other hand, because of specific nature of iron powders, the pressing must be done in multiple steps in order to obtain uniform powder distribution.

The manufacturing cost of SMC cores is directly related to the size of press used for the molding, while the productivity is inversely proportional to the press size. For a given SMC core, it is desirable to choose a smaller press in order to keep the manufacturing cost low. In the case where the volume of the motor is not a big problem, in order to reduce further the manufacturing cost, it is possible to use a low density SMC core. For example, for a 100 ton press, it can produce 500 pieces per hour with a cost of $ 100, resulting in the manufacturing cost of $ 0.2 per piece. For a 500 ton press, on the other hand, it can produce only 100 pieces per hour with a cost of $ 500, and thus the manufacturing cost is $ 5 per piece [34, 69–71]. Therefore, the manufacturing process is a major parameter in design optimization of SMC motors.

B. Assembly process variations

The assembly process variations mainly include the lamination of silicon steel sheets and misplacements of stator, rotor and PMs. Severe misplacements can result in big variations of the motor quality and cause large vibration and excessive resistive torque and mechanical power loss. Table 1.1 lists several manufacturing tolerances (such as magnet strength and skew error) and assembly variations (such as magnet disposition and rotor/stator eccentricity) obtained from industrial motor manufacturing experiences [66].

In the assembling process, due to the structural asymmetry (such as keyway and tag hole), non-uniform material quality (such as the thickness or sand hole) and manufacturing error (such as drill hole and others), and big mechanical disequilibrium in the rotating parts (such as rotors and fans) may appear, and the rotating parts will displace from their gravity centre, resulting in unbalanced centrifugal force and causing the motor to vibrate. Vibration has large negative effects to the motor, such as extra energy consumption, efficiency reduction, direct damage to the

Table 1.1 Variations for some factors of a PM motor [66]

Factor	Description	Ideal	Variation
1	Magnet strength	Nominal	Nominal ± 5 %
2	Skew error	Nominal	Nominal ± 0.6666°
3	Magnetization offset	0°	1.0°
4	Magnet disposition	0°	1.0°
5	Rotor to stator eccentricity	0 mm	0.35 mm

shaft, acceleration of abrasion, which shortens significantly the lifetime of the motor and drive system.

The assembly of stator cores can also be a big challenge for SMC motors. For manufacturing convenience, the stator core of an SMC motor is often molded in separate pieces. Any extra air gap between two pieces of the stator core caused by poor assembly process will result in large reduction of air gap flux density and in turn the motor efficiency. Good assembly structures should be investigated in the design stage of SMC motors.

Figure 1.13 illustrates a multi-disciplinary design framework of electrical motors and drive systems that we propose to take into account the manufacturing quality in mass production in the design stage. The first step is to define the acceptable maximal defect-ratio and system performance. Under these specifications, the motor types, topologies, materials, inverters, and controllers will be designed and optimized under the multi-disciplinary design and machinery technology design.

Due to these manufacturing tolerances, the design optimization of electrical drive systems for mass production is really a challenge in both research and industrial communities as it includes not only the theoretical multi-disciplinary design and analysis but also the practical engineering manufacturing of electrical machines and drive systems. Meanwhile, many new control algorithms, e.g. MPC, have been proposed for motor control, and in the design optimization stage, various many algorithm parameters should be optimized for the best drive system dynamic performance. From the industrial application perspective, it is a natural requirement that the obtained optimal control algorithm parameters are robust against the variations of motor parameters. This is a crucial issue for the batch production of novel drive systems [69–71].

To find effective ways to deal with this problem, several robust design optimization methods have been investigated, such as Taguchi method [72–77] and Six-Sigma robust optimization method [69–71, 78]. These two methods have been found useful to optimize motor performances (including torque ripples, cost, and output power) and quality against the manufacturing tolerances.

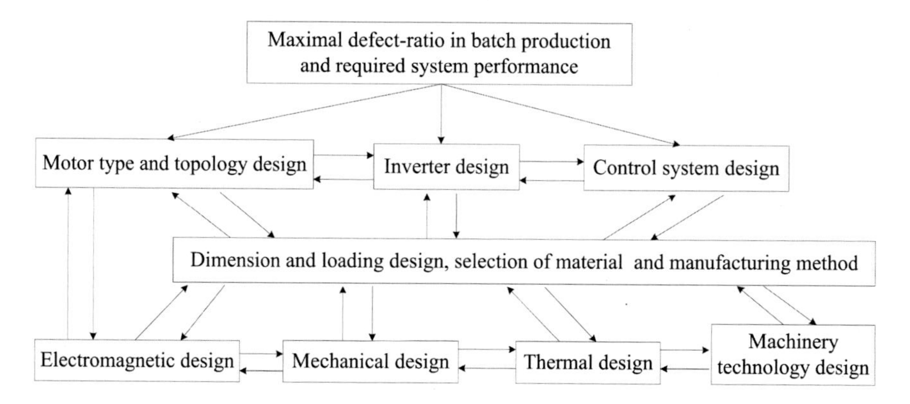

Fig. 1.13 Proposed design framework of drive systems for mass production

In our previous work, several kinds of manufacturing tolerances and assembly variations have been investigated for reducing the cogging torque and harmonics for PM motors, and several dimensional variations have been investigated for optimization of the material cost, output power, and overshoot for an SMC motor and drive system based on a technique known as Design for Six-Sigma (DFSS) [69–71, 78, 79]. DFSS is a robust design technique based on the Six-Sigma methodology. The term Six Sigma is originated from the terminology associated with high quality manufacturing, based on statistical modeling of manufacturing processes. The maturity of a manufacturing process can be described by a sigma rating indicating its yield or the percentage of defect-free products it creates. For the short term quality control, Six-Sigma quality is equivalent to a probability of 99.9999998 %. However, there is about 1.5σ shift from the mean in the long term quality control, so that one six sigma process is actually one in which 99.99966 % of all opportunities to produce some feature of a part are statistically expected to be free of defects (3.4 defects per million opportunities: DPMO) [78, 79]. By taking the DFSS, it can be seen that the motor or system reliabilities can be greatly improved by the proposed methods.

1.4 Major Objectives of the Book

This book presents efficient multi-disciplinary design approaches and application-oriented system-level optimization methods for advanced high quality electrical drive systems. The multi-disciplinary analysis includes materials, electromagnetics, thermotics, mechanics, power electronics, applied mathematics, machinery technology, and quality control and management. This book will benefit the researchers and engineers in the field of design and manufacturing of electric motors and drive systems. The outcomes will enable effective development and high quality mass production of novel high performance drive systems for challenging applications, such as green energy systems and electric vehicles. The main objectives are as follows.

(1) To present a systematic overview of the application-oriented system-level design optimization methods for high quality mass production of advanced electrical drive systems. This is a promising as well as challenging research and application topic in the field of electrical engineering.
(2) To review the popular design optimization methods for electrical machines and drive systems, including design analysis models and methods, such as FEM and magnetic circuit model, and optimization models and algorithms, such as GA and RSM.
(3) To develop novel efficient design optimization methods for electrical machines, including sequential optimization methods for single- and multi-objective problems, multi-level optimization methods for high-dimensional problems and multi-disciplinary design optimization methods for PM machines.

(4) To present system-level design optimization methods for electrical drive systems, which will optimize both the steady-state and dynamic performances of the drive systems, including average output power and material cost of the motor, and overshoot and settling time of the controller.
(5) To develop system-level robust design optimization methods for improving the manufacturing quality of electrical machines and drive systems in mass production.
(6) To present application-oriented design optimization methods for electrical machines and drive systems, including two different applications, where one is a home application, and the other is a HEV application.

1.5 Organization of the Book

Chapter 2 presents an overview of the design fundamentals for electrical machines and drive systems. Design analysis models in terms of different disciplines are investigated, such as the analytical models or methods for electromagnetic and thermal analyses, magnetic circuit model for electromagnetic analysis, FEM for coupled or uncoupled electromagnetic, thermal and mechanical analyses, and FOC and DTC algorithms for the control systems. All these design analysis models can be employed for the performance evaluation of electrical machines and drive systems.

Chapter 3 reviews popular optimization algorithms and approximate models used in optimization of electrical machines as well as electromagnetic devices. The optimization algorithms include the classical gradient-based algorithms and modern intelligent algorithms, such as GA, DEA, and MOGA. Approximate models mainly include RSM, RBF and Kriging models [4–6].

Chapter 4 presents the design optimization methods for electrical machines in terms of different optimization situations, including low- and high-dimension, and single and multi-objectives and disciplines. Five new types of design optimization methods are presented to improve the optimization efficiency of electrical machines, particularly those complex structure PM machines, in terms of different optimization situations.

Chapter 5 aims to present the system-level design optimization methods for electrical drive systems, including single- and multi-level optimization methods. Not only the steady-state but also the dynamic motor performance indicators, such as output power, efficiency, and speed overshoot, are investigated at the same time [1].

It should be noted that the design optimization methods in Chaps. 4 and 5 are under the framework of deterministic approach, which means that all material and structural parameters in the manufacturing process do not have any variations from their nominal values. However, as aforementioned, there are many unavoidable uncertainties or variations in the industrial manufacturing process of electrical machines and drive systems. These variations will affect the reliability and quality of electrical machines and drive systems in mass production, which cannot be investigated by the deterministic approach. Chapter 6 presents a robust design optimization approach based on the technique of DFSS for high quality mass production of high-performance electrical machines and drive systems.

On the other hand, from the perspective of engineering applications, these design optimization methods of electrical machines and drive systems are proposed with several general requirements and constraints, such as the rated torque and given volume and mass, for general applications. Chapter 7 aims to develop application-oriented design optimization methods for electrical machines under deterministic and robust design approaches, respectively. Two kinds of applications are investigated. The first one is about the design optimization of PM-SMC motor for refrigerator and air-conditioner compressors, which can be regarded as home appliance applications. The second one is about the design optimization of flux-switching PM machines for plug-in HEVs drives.

Chapter 8 concludes the whole book and proposes future research and development.

References

1. Waide P, Brunner CU (2011) Energy-efficiency policy opportunities for electric motor-driven systems. International energy agency working paper, energy efficiency series, Paris
2. Industrial efficiency technology database. The Institute for Industrial Productivity (2015). http://ietd.iipnetwork.org/content/motor-systems. Accessed on 20 June 2015
3. Dorrell DG (2014) A review of the methods for improving the efficiency of drive motors to meet IE4 efficiency standards. J Power Electron 14(5):842–851
4. Saidur R (2010) A review on electrical motors energy use and energy savings. Renew Sustain Energy Rev 14:877–898
5. Kumar A, Schei T et al (2011) Hydropower. In: IPCC special report on renewable energy sources and climate change mitigation. Cambridge University Press, Cambridge
6. Islam MR, Lei G, Guo YG, Zhu JG (2014) Optimal design of high frequency magnetic links for power converters used in grid connected renewable energy systems. IEEE Trans Magn 50 (11), Article 2006204
7. Islam MR, Guo YG, Zhu JG (2014) Power converters for medium voltage networks. Springer, Germany
8. Islam MR, Guo YG, Lin ZW, Zhu JG (2014) An amorphous alloy core medium frequency magnetic-link for medium voltage photovoltaic inverters. J Appl Phys 115(17), Article 17E710
9. Islam MR, Guo YG, Zhu JG (2013) A medium frequency transformer with multiple secondary windings for medium voltage converter based wind turbine power generating systems. J Appl Phys 113(17), Article 17A324
10. Xu W, Lei G, Zhang YC, Wang TS, Zhu JG (2012) Development of electrical drive system for the UTS PHEV. In: Proceedings of IEEE energy conversion congress and exposition (ECCE). Raleigh, No. EC-1300. 15–20 Sept 2012
11. Mohammed OA, Lowther DA, Lean MH, Alhalabi B (2001) On the creation of a generalized design optimization environment for electromagnetic devices. IEEE Trans Magn 37(5):3562–3565

12. Fahimi B, Mohammed O (2015) Optimal design of electrical machines. IEEE Trans Energy Convers 30(3):1143
13. Hasanien HM, Abd-Rabou AS, Sakr SM (2010) Design optimization of transverse flux linear motor for weight reduction and performance improvement using RSM and GA. IEEE Trans Energy Conver 25(3):598–605
14. Hasanien HM (2011) Particle swarm design optimization of transverse flux linear motor for weight reduction and improvement of thrust force. IEEE Trans Ind. Electron 58(9):4048–4056
15. Storn R, Price K (1997) Differential evolution- a simple and efficient heuristic for global optimization over continuous spaces. J Global Optim 11:341–359
16. Lei G, Zhu JG, Guo YG, Shao KR, Xu W (2014) Multiobjective sequential design optimization of PM-SMC motors for six sigma quality manufacturing. IEEE Trans Magn 50 (2), Article 7017704
17. Yao D, Ionel DM (2013) A review of recent developments in electrical machine design optimization methods with a permanent magnet synchronous motor benchmark study. IEEE Trans Ind Appl 49(3):1268–1275
18. Ishikawa T, Tsukui Y, Matsunami M (1999) A combined method for the global optimization using radial basis function and deterministic approach. IEEE Trans Magn 35(3):1730–1733
19. Lei G, Yang GY, Shao KR, Guo YG, Zhu JG, Lavers JD (2010) Electromagnetic device design based on RBF models and two new sequential optimization strategies. IEEE Trans Magn 46(8):3181–3184
20. Wang LD, Lowther DA (2006) Selection of approximation models for electromagnetic device optimization. IEEE Trans Magn 42(2):1227–1230
21. Lebensztajn L, Marretto CAR, Costa MC, Coulomb J-L (2004) Kriging: a useful tool for electromagnetic device optimization. IEEE Trans Magn 40(2):1196–1199
22. Mendes MHS, Soares GL, Coulomb J-L, Vasconcelos JA (2013) Appraisal of surrogate modeling techniques: a case study of electromagnetic device. IEEE Trans Magn 49(5):1993–1996
23. Alotto P, Baumgartner U, Freschi F, Köstinger A, Magele Ch, Renhart W, Repetto M (2008) SMES optimization benchmark extended: introducing Pareto optimal solutions into TEAM22. IEEE Trans Magn 44(6):1066–1069
24. Guimaraes FG, Campelo F, Saldanha RR, Igarashi H, Takahashi RHC, Ramirez JA (2006) A multiobjective proposal for the TEAM benchmark problem 22. IEEE Trans Magn 42(4):1471–1474
25. Lebensztajn L, Coulomb JL (2004) TEAM workshop problem 25: A multiobjective analysis. IEEE Trans Magn 40(2):1402–1405
26. Deb K, Pratap A, Agarwal S, Meyarivan T (2002) A fast and elitist multi-objective genetic algorithm: NSGA-II. IEEE Trans Evol Comput 6(2):182–197
27. Reyes-Sierra M, Coello CAC (2006) Multi-objective particle swarm optimizers: A survey of the state-of-the-art. Int J Comput Intell Res 2(3):287–308
28. Dias AHF, Vasconcelos JA (2002) Multiobjective genetic algorithms applied to solve optimization problems. IEEE Trans Magn 38(2):1133–1136
29. dos Santos Coelho L, Alotto P (2008) Multiobjective electromagnetic optimization based on a nondominated sorting genetic approach with a chaotic crossover operator. IEEE Trans Magn 44(6):1078–1081
30. Ho SL, Yang SY, Ni GZ, Lo EWC, Wong HC (2005) A particle swarm optimization-based method for multiobjective design optimizations. IEEE Trans Magn 41(5):1756–1759
31. Xie DX, Sun XW, Bai BD, Yang SY (2008) Multiobjective optimization based on response surface model and its application to engineering shape design. IEEE Trans Magn 44(6):1006–1009
32. Di Barba P (2010) Multiobjective shape design in electricity and magnetism. Lecture notes in electrical engineering, vol. 47
33. Lei G, Shao KR, Guo YG, Zhu JG (2012) Multi-objective sequential optimization method for the design of industrial electromagnetic devices. IEEE Trans Magn 48(11):4538–4541

34. Lei G, Guo YG, Zhu JG et al (2015) Techniques for multilevel design optimization of permanent magnet motors. IEEE Trans Energy Conver 30(4):1574–1584
35. Lei G, Guo YG, Zhu JG, Chen XM, Xu W (2012) Sequential subspace optimization method for electromagnetic devices design with orthogonal design technique. IEEE Trans Magn 48 (2):479–482
36. Lei G, Xu W, Hu JF, Zhu JG, Guo YG. Shao KR (2014) Multilevel design optimization of a FSPMM drive system by using sequential subspace optimization method. IEEE Trans Magn 50(2), Article 7016904
37. Wang SH, Meng XJ, Guo NN, Li HB, Qiu J, Zhu JG et al (2009) Multilevel optimization for surface mounted PM machine incorporating with FEM. IEEE Trans Magn 45(10):4700–4703
38. Lei G, Liu CC, Guo YG, Zhu JG (2015) Multidisciplinary design analysis for PM motors with soft magnetic composite cores. IEEE Trans Magn 51(11), Article 8109704
39. Lei G, Wang TS, Guo YG, Zhu JG, Wang SH (2014) System level design optimization methods for electrical drive systems: deterministic approach. IEEE Trans Ind Electron 61 (12):6591–6602
40. Ramsden VS, Dunlop JB, Holliday WM (1992) Design of a hand-held motor using a rare earth permanent magnet rotor and glassy metal stator. In: Proceedings of international conference on electrical machines, pp 376–380
41. Ramsden VS, Watterson PA, Dunlop JB (1992) Optimization of REPM motors for specific applications (invited). In: Proceedings of the international workshop on rare earth materials and applications, Canberra, pp 296–307
42. Ramsden VS et al (1992) Optimization of a submersible brushless rare-earth permanent-magnet motor for solar power. In: Proceedings of internationl workshop on electric and magnetic fields, Liege, pp 461–464
43. Lovatt HC, Ramsden VS, Mecrow BC (1997) Design of an in-wheel motor for a solar-powered electric vehicle. In: Proceedings of international conference on electrical machines and drives, pp 234–238
44. Ramsden VS, Zhu JG et al (2001) High performance electric machines for renewable energy generation and efficient drives. J Renew Energy 22(1–3):159–167
45. Widdowson GP, Howe D, Evison PR (1991) Computer-aided optimization of rare-earth permanent magnet actuators. In: Proceedings of international conference on computation in electromagnetics, 25–27 Nov 1991, pp 93–96
46. Chhaya SM, Bose BK (1993) Expert system based automated simulation and design optimization of a voltage-fed inverter for induction motor drive. Int Conf IECI 2:1065–1070, 15–19 Nov 1993
47. Guo YG, Zhu JG, Liu DK, Wang SH (2007) Application of multi-level mult-domain modelling in the design and analysis of a PM transverse flux motor with SMC core. In: Proceedings of the 7th international conference on power electronics and drive systems (PEDS07), Bangkok, Thailand, pp 27–31
48. Lei G, Liu CC, Zhu JG, Guo YG (in press) Robust multidisciplinary design optimization of PM machines with soft magnetic composite cores for batch production. IEEE Trans Magn (in press)
49. Guo YG, Zhu JG, Dorrell D (2009) Design and analysis of a claw pole PM motor with molded SMC core. IEEE Trans Magn 45(10):582–4585
50. Zhu JG, Guo YG, Lin ZW, Li YJ, Huang YK (2011) Development of PM transverse flux motors with soft magnetic composite cores. IEEE Trans Magn 47(10):4376–4383
51. Guo YG, Zhu JG, Watterson PA, Wei Wu (2006) Development of a PM transverse flux motor with soft magnetic composite core. IEEE Trans Energy Conver 21(2):426–434
52. Huang YK, Zhu JG et al (2009) Thermal analysis of high-speed SMC motor based on thermal network and 3D FEA with rotational core loss included. IEEE Trans Magn 45(106):4680–4683
53. Liu CC, Zhu JG, Wang YH, Guo YG, Lei G, Liu XJ (2015) Development of a low-cost double rotor axial flux motor with soft magnetic composite and ferrite permanent magnet materials. J. Appl Phys 117(17), Article# 17B507

54. Zhu ZQ, Howe D (2007) Electrical machines and drives for electric, hybrid, and fuel cell vehicles. Proc IEEE 95(4):746–765
55. Emadi A, Lee YJ, Rajashekara K (2008) Power electronics and motor drives in electric, hybrid electric, and plug-in hybrid electric vehicles. IEEE Trans Ind Electron 55(6):2237–2245
56. Kreuawan S, Gillon F, Brochet P (2008) Optimal design of permanent magnet motor using multidisciplinary design optimization. In: Proceeings of 18th international conference on electrical machines, Vilamoura, 6–9 Sept 2008, pp 1–6
57. Vese I, Marignetti F, Radulescu MM (2010) Multiphysics approach to numerical modeling of a permanent-magnet tubular linear motor. IEEE Trans Ind Electron 57(1):320–326
58. Buja GS, Kazmierkowski MP (2004) Direct torque control of PWM inverter-fed AC motors—a survey. IEEE Trans Ind Electron 51(4):744–757
59. Kouro S, Cortes P, Vargas R, Ammann U, Rodriguez J (2009) Model predictive control—a simple and powerful method to control power converters. IEEE Trans Ind Electron 56 (6):1826–1838
60. Bolognani S, Bolognani S, Peretti L, Zigliotto M (2009) Design and implementation of model predictive control for electrical motor drives. IEEE Trans Ind Electron 56(6):1925–1936
61. Xia CL, Wang YF, Shi TN (2013) Implementation of finite-state model predictive control for commutation torque ripple minimization of permanent-magnet brushless DC motor. IEEE Trans Ind Electron 60(3):896–905
62. Morel F, Lin-Shi XF, Retif J-M, Allard B, Buttay C (2009) A comparative study of predictive current control schemes for a permanent-magnet synchronous machine drive. IEEE Trans Ind Electron 56(7):2715–2728
63. Zhang YC, Zhu JG, Xu W, Guo YG (2011) A simple method to reduce torque ripple in direct torque-controlled permanent-magnet synchronous motor by using vectors with variable amplitude and angle. IEEE Trans Ind Electron 58(7):2848–2859
64. Wang TS, Zhu JG, Zhang YC (2011) Model predictive torque control for PMSM with duty ratio optimization. In: Proceedings of ICEMS, 20–23 Aug 2011, pp 1-5
65. Liu C-H, Hsu Y-Y (2010) Design of a self-tuning PI controller for a STATCOM using particle swarm optimization. IEEE Trans Ind Electron 57(2):702–715
66. Khan MA, Husain I, Islam MR, Klass JT (2014) Design of experiments to address manufacturing tolerances and process variations influencing cogging torque and back EMF in the mass production of the permanent-magnet synchronous motors. IEEE Trans Ind Appl 50 (1):346–355
67. Coenen I, Giet M, Hameyer K (2011) Manufacturing tolerances: Estimation and prediction of cogging torque influenced by magnetization faults. In: Proceedings of 14th European conference on power electronics and applications, pp 1–9
68. Gasparin L, Fiser R (2011) Impact of manufacturing imperfections on cogging torque level in PMSM. In: Proceedings of 2011 IEEE ninth international conference on power electronics and drive systems (PEDS), pp 1055–1060
69. Lei G, Wang TS, Zhu JG, Guo YG, Wang SH (2015) System level design optimization method for electrical drive system: robust approach. IEEE Trans Ind Electron 62(8):4702–4713
70. Lei G, Guo YG, Zhu JG et al (2012) System level six sigma robust optimization of a drive system with PM transverse flux machine. IEEE Trans Magn 48(2):923–926
71. Lei G, Zhu JG, Guo YG, Hu JF, Xu W, Shao KR (2013) Robust design optimization of PM-SMC motors for Six Sigma quality manufacturing. IEEE Trans Magn 49(7):3953–3956
72. Taguchi G, Chowdhury S, Wu Y (2004) Taguchi's quality engineering handbook, Wiley
73. Omekanda AM (2006) Robust torque and torque-per-inertia optimization of a switched reluctance motor using the Taguchi methods. IEEE Trans Ind Appl 42(2):473–478
74. Kim S-I, Lee J-Y, Kim Y-K et al (2005) Optimization for reduction of torque ripple in interior permanent magnet motor by using the Taguchi method. IEEE Trans Magn 41(5):1796–1799
75. Ashabani M, Mohamed YA, Milimonfared J (2010) Optimum design of tubular permanent-magnet motors for thrust characteristics improvement by combined Taguchi-neural network approach. IEEE Trans Magn 46(12):4092–4100

76. Hwang C-C, Lyu L-Y, Liu C-T, Li P-L (2008) Optimal design of an SPM motor using genetic algorithms and Taguchi method. IEEE Trans Magn 44(11):4325–4328
77. Lee S, Kim K, Cho S et al (2014) Optimal design of interior permanent magnet synchronous motor considering the manufacturing tolerances using Taguchi robust design. IET Electr Power Appl 8(1):23–28
78. Koch PN, Yang RJ, Gu L (2004) Design for six sigma through robust optimization. Struct Multidiscipl Optim 26(3–4):235–248
79. Meng XJ, Wang SH, Qiu J, Zhang QH, Zhu JG, Guo YG, Liu DK (2011) Robust multilevel optimization of PMSM using design for six sigma. IEEE Trans Magn 47(10):3248–3251

Chapter 2
Design Fundamentals of Electrical Machines and Drive Systems

Abstract This chapter presents a brief summary of the design fundamentals including the analysis models and methods for electrical machines and drive systems, based on our design experiences, particularly for permanent magnet electrical machine with soft magnetic composite cores. Because of the multi-disciplinary nature, these design models and methods will be investigated at the disciplinary level, including electromagnetic, thermal, mechanical, power electronics, and control algorithm designs. Several design examples will be presented to illustrate the corresponding design models and methods based on our research findings, such as the finite element model for design analysis of motors, and the model predictive control algorithm and its improvement form for the drive systems. These models and algorithms will be employed in the design optimization of electrical machines and drive systems in the following chapters.

Keywords Electrical drive systems · Electromagnetic design · Thermal design · Mechanical design · Power electronics design · Control algorithms · Finite element model · Model predictive control

2.1 Introduction

2.1.1 Framework of Multi-disciplinary Design

Figure 2.1 illustrates a general framework of multi-disciplinary design for electrical machines and drive systems. As shown, three main components, i.e., motor, power electronics and controller, have to be investigated when designing such electrical drive systems [1, 2]. The main design procedure includes the following steps.

Firstly, define the specifications of the electrical machine and drive system required by a given application, which include the steady state specifications, such as the rated power, speed range, voltage, current, efficiency, power factor (in case of AC machines), volume and cost, and dynamic performances, such as the maximum overshoot, settling time, and stability.

G. Lei et al., *Multidisciplinary Design Optimization Methods for Electrical Machines and Drive Systems*, Power Systems,
DOI 10.1007/978-3-662-49271-0_2

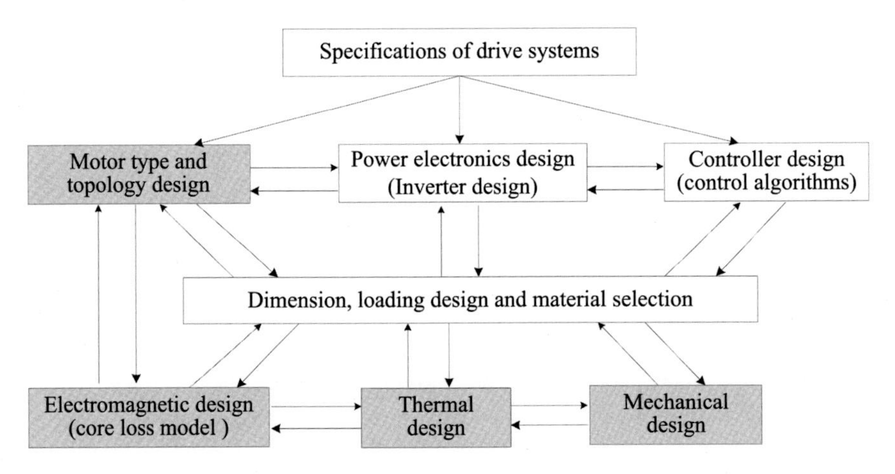

Fig. 2.1 Multi-disciplinary design framework of electrical machines and drive systems

Secondly, select a type of the motor, power electronic converter, and control algorithm from possible options. The motor options include permanent magnet (PM) motors, induction machines, synchronous machines, DC machines, and switched reluctance machines. For servo drives, stepping motors and other types of servo motors can be considered. In this step, different motor topologies have to be investigated as well. The power electronic converter options mainly include the different topologies of AC/DC, DC/DC, and DC/AC converters. The controller design mainly investigates the control strategies and algorithms, such as field oriented control (FOC), direct torque control (DTC), and model predictive control (MPC).

Thirdly, based on the selected motor type, converter circuit, and control scheme, various disciplinary-level analyses should be conducted to evaluate the performance of the drive system. For example, the motor design analysis consists of mainly the electromagnetic, thermal and mechanical analyses (the shaded boxes in the figure). Coupled-field analyses may be required in the design process, such as electromagnetic-thermal and electromagnetic-mechanical stress analyses.

In summary, the design of electrical machines and drive systems mainly consists of the analyses of five coupled disciplines or domains: electromagnetic, thermal, mechanical, power electronics, and controller designs. The following sections will present the popular design analysis models and methods for each discipline.

2.1.2 Power Losses and Efficiency

Power losses and efficiency are two main issues in the design analysis of electrical machines and drive systems. The power losses are mainly composed of the copper loss, core loss, mechanical loss, and stray loss.

(1) The copper loss or Ohmic loss: $P_{Cu} = I^2R$ is the power dissipated in stator and rotor windings due to the resistance of copper wire, where I is the winding current and R the winding resistance. Normally the DC resistance is used in the calculation. However, it should be noted that the winding resistance depends on the operating conditions, i.e., temperature and frequency (due to the skin effects). In case of the brushes and slip rings/commutator, the effect of contact resistance is often accounted for by assuming a voltage drop of 2 V.

(2) The core loss is the power dissipated in a magnetic core due to the variation of magnetic field. This occurs in the stator and/or rotor iron core of an electrical machine subject to AC excitations. Practically, it can be measured by open-circuit or no-load tests. When the magnetic material is under an alternating sinusoidal flux excitation, the alternating core loss can be calculated by

$$P_a = C_{ha} f B^h + C_{ea}(fB)^2 + C_{aa}(fB)^{1.5} \tag{2.1}$$

where f is the excitation frequency, B the magnitude of sinusoidal magnetic flux density, and C_{ha}, C_{ea}, C_{aa}, and h are the alternating core loss coefficients. In case of rotating electrical machines, the rotational core losses have to be considered. Figure 2.2 plots the average core losses with alternating flux density from 2 to 2,000 Hz and circular rotating flux density vectors from 5 to 1,000 Hz of a cubic soft magnetic composite (SMC) SOMALY™ 500 sample [3]. These are the standard core loss data used to identify the core loss model parameters. The circularly rotational core loss can be calculated by

$$P_r = P_{hr} + C_{er}(fB)^2 + C_{ar}(fB)^{1.5} \tag{2.2}$$

where

$$\frac{P_{hr}}{f} = a_1\left[\frac{1/s}{(a_2 + 1/s)^2 + a_3^2} - \frac{1/(2-s)}{[a_2 + 1/(2-s)]^2 + a_3^2}\right]$$

and

$$s = 1 - \frac{B}{B_s}\sqrt{1 - [1/(a_2^2 + a_3^2)]}$$

B_s is the saturation flux density, and C_{er}, C_{ar}, a_1, a_2 and a_3 are the rotational core loss coefficients.

When the material is under a two dimensional elliptically rotating **B** excitation, the core loss can be computed by

$$P_{er} = R_B P_r + (1 - R_B)^2 P_a \tag{2.3}$$

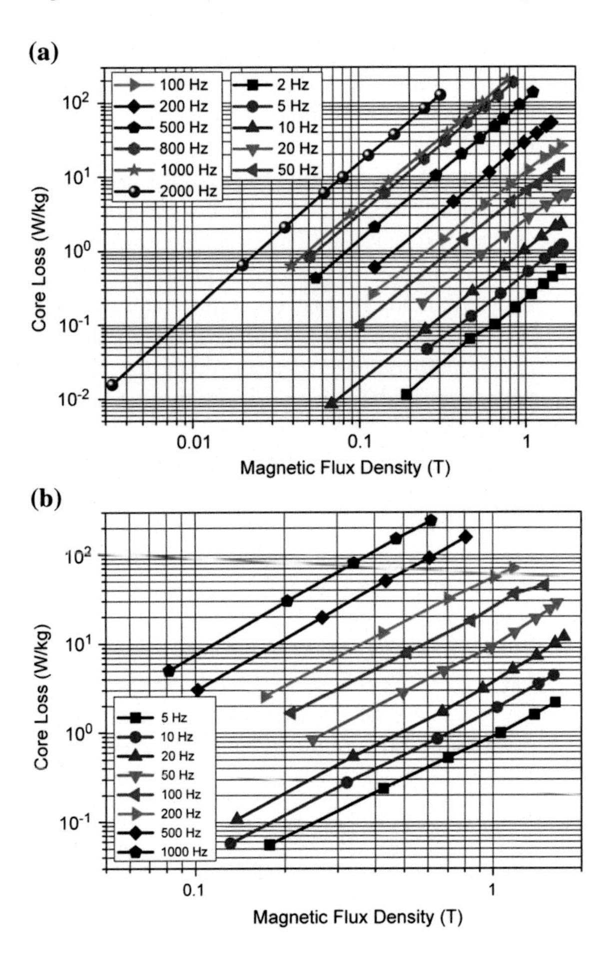

Fig. 2.2 Average core losses under **a** alternating and **b** circular rotating magnetic fluxes [3]

where $R_B = B_{\min}/B_{\text{maj}}$ is the axis ratio, $B_{\min}$ and B_{maj} are the magnitudes of the minor and major axes of the ellipse, respectively, and P_r and P_a the corresponding rotational and alternating core losses when $B = B_{\text{maj}}$. More details about the rotational core losses can be found in [3–9].

(3) The mechanical losses are the power losses caused by the friction (brushes, slip rings/commutator, shaft and bearing), damping, windage, and cooling fan. It can be approximately determined by no-load test. In design, empirical data are used.

(4) The stray loss is the power loss caused by stray factors that are hard to determine separately, such as the non-uniform current distribution in conductors and additional core loss due to distorted magnetic flux distribution for various reasons. Because it is usually difficult to determine accurately the stray loss, estimations based on experimental tests and empirical judgment are

acceptable. For most types of machines, this can be assumed to be 1 % of the output power.

In most electrical machines, the stator and/or rotor cores subject to varying magnetic fluxes are made of laminated silicon steels, which have low core loss, and hence the major power loss is the copper loss. Depending on the type of machine, the copper loss normally accounts for 80–90 % of the total loss.

Based on the above analysis, the efficiency of a machine can be calculated by

$$\eta = \frac{P_{out}}{P_{in}} = \frac{P_{in} - (P_{Cu} + P_{Core} + P_{Mec} + P_{Stray})}{P_{in}} \tag{2.4}$$

Typical values of full load efficiency for rotating machines are:

- 50 % or less for fractional horse power motors (a few W to a few hundreds of W),
- 75–85 % for electrical machines of 1 kW to a few tens of kW,
- 85–95 % for electrical machines of 100 kW to 1 MW, and
- 95–98 % or above for electrical machines of 1 MW to a few hundreds of MW (e.g. 98 % for 100 MVA turbo generator).

2.2 Electromagnetic Design

Since electrical machines are electromagnetic devices for transforming electrical power at one voltage to another (transformers) or converting electric power into mechanical power or vice versa (motors or generators) by the principle of electromagnetic induction, electromagnetic design is a fundamental design stage of electrical machines and drive systems, and is usually based on the following three kinds of analysis models: the analytical model, magnetic circuit model, and finite element model (FEM) [10–20].

2.2.1 Analytical Model

Analytical model is generally used to calculate the performance indicators of electrical machines, such as the output power, torque, and cogging torque. For example, the power and sizing equations are the powerful ways to guide the design of PM motors [10, 11]. By utilizing the current density in the sizing equation, some basic internal relationships can be found among the main dimensions to maximize the torque density.

Assuming the flux linkage of stator winding in a PM motor is sinusoidal, and ignoring the winding resistance, the input power P_{in} can be expressed by

$$P_{in} = \frac{m}{T}\int_{0}^{T} e(t)i(t)dt = \frac{m}{T}\int_{0}^{T} E_m sin\left(\frac{2\pi}{T}t\right) I_m sin\left(\frac{2\pi}{T}t\right) dt = \frac{m}{2}E_m I_m \tag{2.5}$$

where m is the number of phases, E_m the peak value of back electromotive force (EMF), I_m the peak value of phase current, and T the electrical time period. The output torque can be calculated by

$$T_{out} = \frac{P_{out}}{\omega_r} = \eta \frac{m}{2} p \lambda_p K_{sf} A_s J_m \tag{2.6}$$

where η is the efficiency, p the number of pole pairs, λ_p the peak value of PM flux linkage, ω_r the mechanical rotary speed, K_{sf} the slot fill factor, A_s the slot area, and J_m the peak of current density. For different kinds of PM motors, λ_p and A_s are related differently to their dimensions [21–23].

2.2.2 Magnetic Circuit Model

The magnetic circuit model acts as a uniform principle in descriptive magnetostatics, and as an approximate computational aid in electrical machine design. The model uses the conception of magnetic reluctance to establish an equivalent circuit for approximate analysis of static magnetic field in electrical machines [24]. To illustrate this model, a PM transverse flux machine (TFM) designed by SMC material is investigated.

The SMC material is a relatively new soft magnetic material that has many advantages over the conventional silicon steel sheets. The main advantages of SMC material are the magnetic and mechanical isotropy and low cost, high productivity, and high quality manufacturing capability of complex electromagnetic components by the matured powder metallurgical molding technology, which will enable low cost high productivity commercial manufacturing of SMC motors for a great variety of electrical appliances [24–32].

In our previous work, a 3D flux PM TFM with SMC stator core was developed. Figure 2.3 shows a photo of the PM-SMC TFM prototype. This machine was initially designed to deliver an output power of 640 W at 1800 rev/min. It has 20 poles in the external PM rotor, i.e., 120 PMs in the rotor and 60 SMC teeth in the stator. The stator core is made of SMC SOMALOY™ 500. The operating frequency of this motor is 300 Hz at 1800 rev/min. Table 2.1 tabulates the main design dimensions for this TFM [24, 25].

In order to briefly predict the performance of this TFM, a sketchy magnetic circuit model as shown in Fig. 2.4 can be used. Figure 2.4 illustrates the main flux circuit model and flux path of this TFM, where the resistances represent the magnetic reluctances and the current source (PhiPM) stands for the magneto-motive forces (mmfs) of PMs, and thus R_{ry} represents the magnetic reluctance of rotor, R_m the

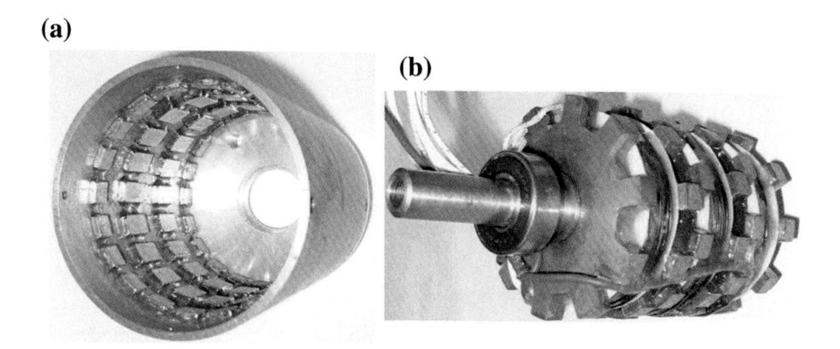

Fig. 2.3 Photo of the PM-SMC TFM prototype, **a** PM rotor, and **b** 3 stack SMC stator

Table 2.1 Main design dimensions of PM-SMC TFM

Par.	Description	Unit	Value
–	Number of phases	–	3
–	Number of poles	–	20
–	Number of stator teeth	–	60
–	Number of magnets	–	120
–	Stator outer radius	mm	40
–	Effective stator axial length	mm	93
x_1	PM circumferential angle	degree	12
x_2	PM width	mm	9
x_3	SMC tooth circumferential width	mm	9
x_4	SMC tooth axial width	mm	8
x_5	SMC tooth radial height	mm	10.5
x_6	Number of turns	–	125
x_7	Diameter of copper wire	mm	1.25
x_8	Air gap length	mm	1.0

magnetic reluctance of PM, R_g the magnetic reluctance of the air gap, R_{st1} the magnetic reluctance of the stator teeth, R_{st2} and R_{sy} stand for the magnetic reluctance of the stator yoke. By analyzing this model, the main magnetic flux can be calculated.

Meanwhile, the magnetic flux leakage is a serious problem in this TFM, thus it should be considered in the magnetic circuit model. Several flux leakage models can be constructed for this TFM. Figure 2.5 illustrates the main flux leakage model. In this model, the adjacent PM in the one side of the machine is modeled, where R_{ry1} represents the magnetic reluctance of rotor, R_{g1} and R_{g2} represent the magnetic reluctance of the air gap, R_{s1} stands for the magnetic reluctance of the stator.

With the computed flux linkage, the resultant magnetic flux density in the air gap and the flux per turn of coil can be estimated. After calculation, the obtained flux per turn of this PM-SMC TFM is 0.32 mWb, which is higher than the calculated result (0.28 mWb) by using the FEM [24].

This model can be also used to evaluate the performance of the motor. Based on the calculated magnetic flux of the motor, the flux linkage per phase equals the

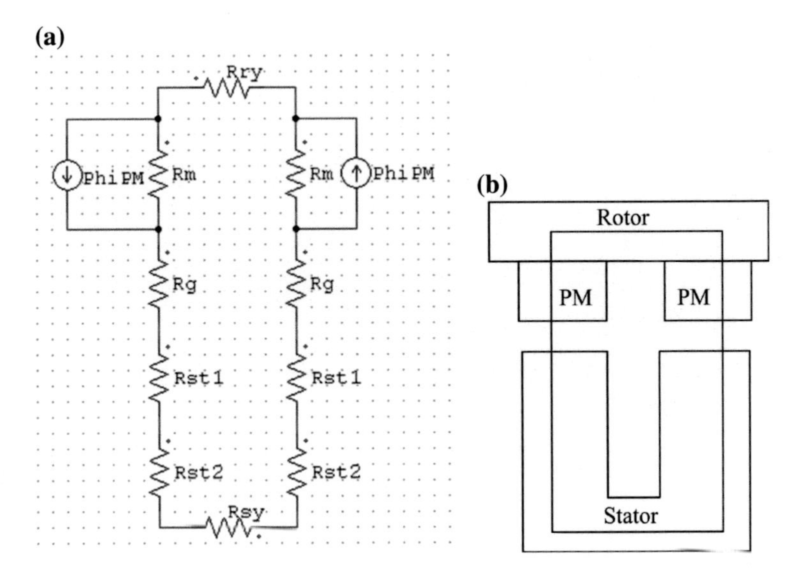

Fig. 2.4 Main flux circuit and flux path of the PM-SMC TFM, **a** magnetic circuit model, **b** flux path in 2D plane

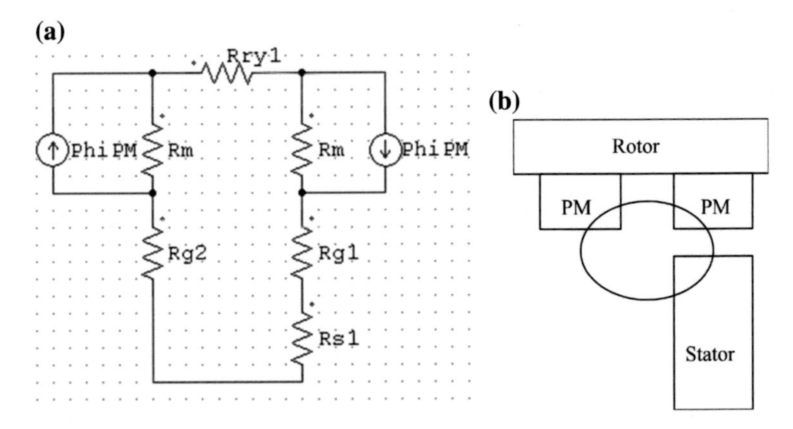

Fig. 2.5 Flux leakage circuit and path of the PM-SMC TFM, **a** magnetic circuit model, **b** leakage path in 2D plane

number of coil turns multiplied by the magnetic flux of each coil turn, and it can be computed as

$$\lambda_{PM} = k_l N_{coil} p \Phi_{gap} \tag{2.7}$$

where λ_{PM} is the PM flux linkage per phase, k_l the leakage coefficient, N_{coil} the number of turns of the phase winding, p the number of pole pairs, and Φ_{gap} is flux per coil turn. The back EMF can be expressed as

$$E_m = \omega_e \lambda_{PM} = p\omega_m \lambda_{PM} \tag{2.8}$$

where $\omega_e = p\omega_m$ is the electrical angular frequency, and ω_m the mechanical angular speed. The electromagnetic torque T_{em} can be expressed as

$$T_{em} = \frac{P_{em}}{\omega_m} = \frac{\sqrt{2}}{2} mp\lambda_{PM} I_m \tag{2.9}$$

After the calculation, the no-load back EMF is 53.26 V at the rated speed of 1800 rev/min. According to (2.9), the electromagnetic torque is 4.66 Nm at the rated current of 5.5 A (RMS value). Compared to the electromagnetic torque obtained from FEM, i.e., 4.08 Nm, the relative error is about 0.58/4.08 = 14.21 %.

2.2.3 *Finite Element Model*

FEM is a widely used analysis model for field analysis in electrical machines as well as other electromagnetic devices. The theory of FEM can be found in many books and research papers. The PM-SMC TFM investigated above will be employed as an example to show the application of FEM for designing electrical machines.

When analyzing the magnetic field distribution, we used field analysis software package ANSYS, and taking advantage of the periodical symmetry, we only need to analyze one pole-pair region of the machine, as shown in Fig. 2.6a. At the two radial boundary planes, the magnetic scalar potential obeys the periodical boundary conditions:

$$\varphi_m(r, \Delta\theta, z) = \varphi_m(r, -\Delta\theta, z) \tag{2.10}$$

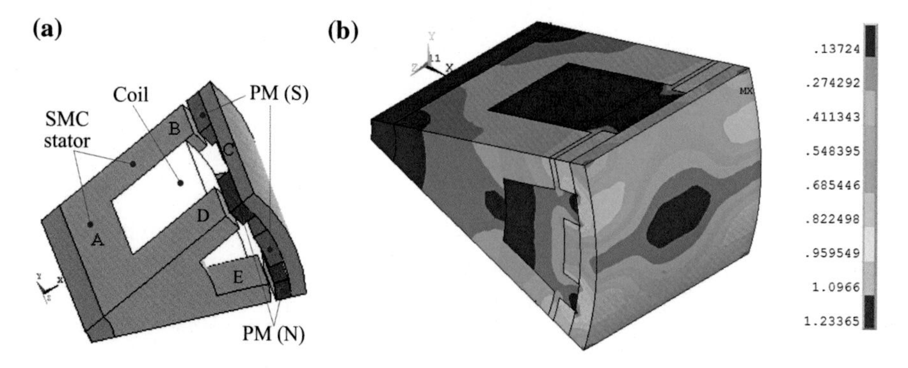

Fig. 2.6 **a** One pole pitch of FEM solution region for one phase (stack), and **b** magnetic field distribution under no-load

Table 2.2 Key PM-SMC TFM parameters

Parameter	Unit	Calculated	Measured
Motor back EMF constant	Vs	0.247	0.244
Phase resistance	Ω	0.310	0.305
Phase inductance	mH	6.68	6.53
Maximal cogging torque	Nm	0.339	0.320

where $\Delta\theta = 18°$ mechanical is the angle of one pole pitch. The origin of the cylindrical coordinate is located at the center of the stack.

Figure 2.6b illustrates the magnetic field distribution under no-load. Based on the FEM analysis, the calculated key motor parameters for this machine are listed in Table 2.2. The measured parameters are also listed in the table to show the effectiveness of the FEM method. As shown, the measured motor back EMF constant is 0.244 Vs, 1 % lower than the calculated value of 0.247 Vs. The calculated phase resistance and inductance, and maximal cogging torque are 0.310 Ω, 6.68 mH and 0.339 Nm, respectively, which are very close to the measured values (0.305 Ω, 6.53 mH and 0.320 Nm). In summary, the estimated parameters calculated by the FEM-based method are well aligned with the experimental results. Therefore, FEM is better than magnetic circuit model, and it is reliable to be used for optimization of the electromagnetic design of electrical machines.

Moreover, the output performance parameters, such as output power, torque and efficiency, can be estimated with the calculated electromagnetic parameters mentioned above. In the estimation, the control method is assumed to maintain that the *d*-axis component of current equals zero. Figure 2.7 shows the per phase equivalent electric circuit of this motor under the assumed control method.

Based on this per phase equivalent electrical circuit, the main relationships of the motor can be predicted by

$$V_{in} = \sqrt{(E_a + I_a R_a)^2 + (\omega_e L_a I_a)^2} \tag{2.11}$$

$$P_{in} = 3 V_{in} I_a \cos\varphi \tag{2.12}$$

$$P_{out} = P_{in} - P_{core} - P_{copper} - P_{mech} \tag{2.13}$$

$$T_{out} = \frac{P_{out}}{\omega_r} \tag{2.14}$$

where V_{in} is the input voltage, E_a the back EMF, I_a the armature current, ω_e the electric angular frequency, L_a the inductance, R_a the resistance, φ the angle between V_{in} and E_a, P_{in} the input power, P_{out} the output power, P_{core} the core loss, P_{copper} the copper loss, P_{mech} the mechanical loss, T_{out} the output torque, and ω_r the mechanical angular speed.

In motor with SMC cores, unlike the conventional motors made of silicon sheet steels, the core loss can be a major part among all power losses, and the mechanical

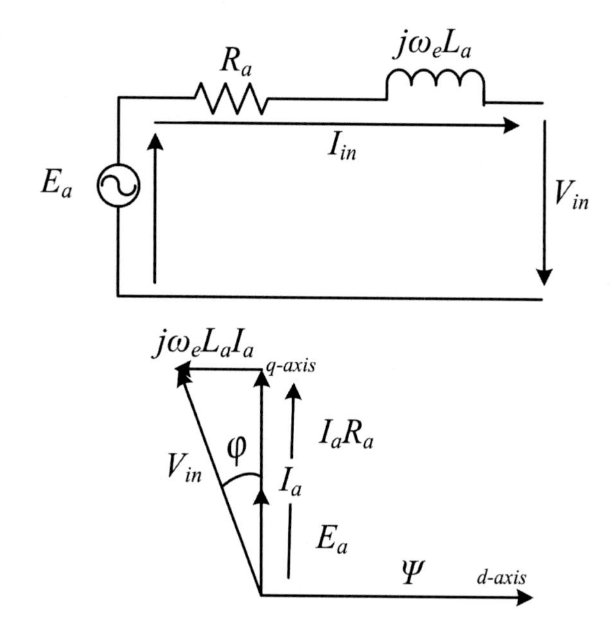

Fig. 2.7 Per phase equivalent electric circuit and phasor diagrams of the motor

loss is generally considered as 1–1.5 % of the output power. In general, the core loss prediction in the TFM should be calculated by using the FEM based on the multi-frequency core loss characteristic of the material. More comparison results can be seen in [24, 25].

2.3 Thermal Design

2.3.1 Thermal Limits in Electrical Machines

The rating of an electrical machine gives its working capability under the specified electrical and environmental conditions. Major factors that determine the ratings are thermal and mechanical considerations. To obtain an economic utilization of the materials and safe operation of the motor, it is necessary to predict with reasonable accuracy the temperature rise of the internal parts, especially in the coils and magnets.

The temperature rise resulted from the power losses in an electrical machine plays a key role in rating the power capacity of the machine, i.e., the amount of power it can convert without being burnt for a specified length of life time. The life expectancy of a large industrial electrical machine ranges from 10 to 50 years or more. In an aircraft or electronic equipment, it can be of the order of a few thousand hours, whereas in a military application, e.g. missile, it can be only a few minutes.

Table 2.3 Classification of electrical insulation materials

Class	Maximum temperature rise (°C)	Materials
O	90	Paper, cotton, silk
A	105	Cellulose, phenolic resins
B	130	Mica, glass, asbestos with organic binder
F	155	Same as above with suitable binder
H	180	Mica, glass, asbestos with silicone binder, silicone resin, Teflon

The operating temperature of a machine is closely associated with its life expectancy because deterioration of the insulation is a function of both time and temperature. Such deterioration is a chemical phenomenon involving slow oxidation and brittle hardening, leading to loss of mechanical durability and dielectric strength. In many cases the deterioration rate is such that the life of the insulation can be expressed as

$$Life\_Time = Ae^{B/T} \tag{2.15}$$

where A and B are constants and T is the absolute temperature. Roughly, it says that for each 10 °C temperature rise exceeding the maximum allowable temperature rise, the life time of insulation is halved.

Insulation materials used in electrical machines are classified by the maximum allowable temperature rise that can be safely withstood. Table 2.3 lists the classification of electrical insulation materials by the IEC (International Electrotechnical Commission).

Generally, there are two kinds of analysis models for thermal analysis in electrical machines, namely the thermal network model and the FEM [14, 15, 20, 25]. The following sections will present examples for the two methods.

2.3.2 *Thermal Network Model*

Two design examples will be illustrated to show the usage of thermal network model for the thermal analysis of PM-SMC motors. The first one is a TFM, and the second is a high speed claw pole motor.

A. Transverse flux machine

In this study, the temperature rise was calculated by using a hybrid thermal network model with distributed heat sources, as shown in Fig. 2.8.

For high computation accuracy, every part, e.g. the air gap, is divided into two or more segments. The thermal resistances to heat conduction in the following sections are calculated: rotor yoke (R_{ry}), magnets (R_m), glue between magnets and rotor yoke

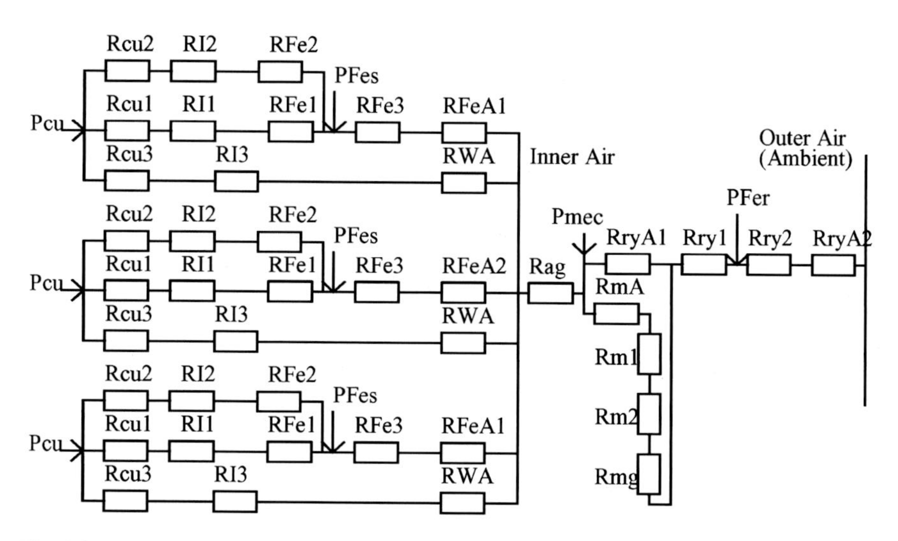

Fig. 2.8 Thermal network model of the TFM prototype

(R_{mg}), air gap (R_{ag}), stator yoke (R_{Fe1}), stator side discs (R_{Fe2}), stator teeth (R_{Fe3}), varnished copper wire (R_{cu}), and insulations (R_{I1}, R_{I2}, R_{I3}) between the winding and the stator yoke, the stator wall disc, and the air gap, respectively. In addition, the thermal resistances of the stator shaft (R_{ss}), the aluminum end plates (R_{al}), and the stationary air (R_{sa}) between the side stator discs and the end plates are calculated separately [25].

The equivalent thermal resistances to the heat convection of the following sections are calculated: that between the stator tooth surface and the inner air in the air gap (R_{FeA}), that between the winding and the inner air (R_{WA}), that between the magnet and the inner air (R_{mA}), that between the rotor yoke and the inner air (R_{ryA1}), and that between the rotor yoke and the outer air (R_{ryA2}).

The heat sources include the stator winding copper losses (P_{cu}), the stator and rotor core losses (P_{Fes}, P_{Fer}), and the mechanical losses due to windage and friction (P_{mec}). The improved method for core loss calculation can obtain the loss distribution, which is a great advantage for thermal calculation by the hybrid thermal model.

The temperature rises in the middle of several parts are calculated as 64.9 °C in the stator winding, 78.6 °C in the stator core, 59.3 °C in the air gap, 36.1 °C in the magnets, and 25.3 °C in the rotor yoke outer surface. The experimentally measured results are 66 °C in the stator winding and 27 °C in the rotor yoke, and it can be seen that the maximum relative error between the calculated and measured results is only 3 %. Thus, it is reliable to use the thermal network method for design of this TFM.

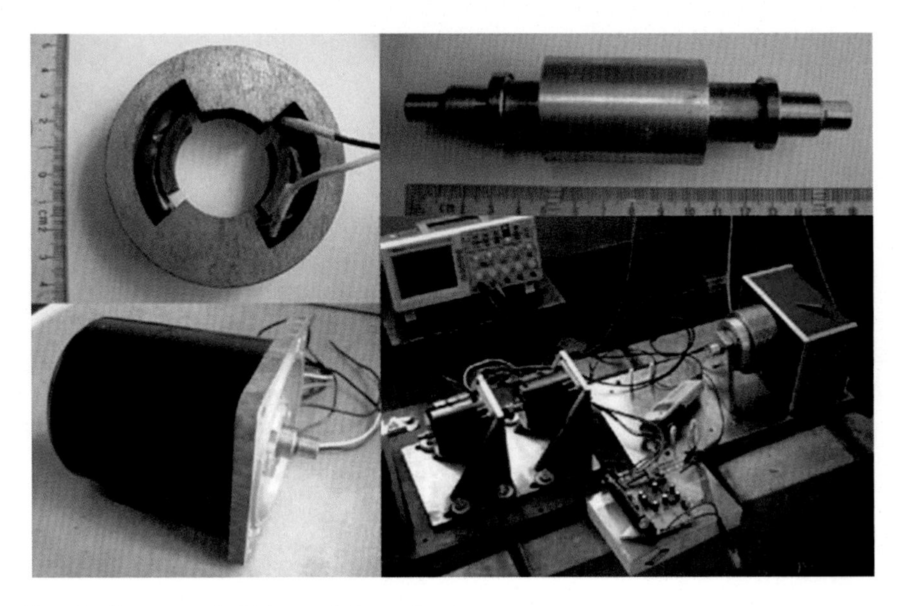

Fig. 2.9 Prototype of a high speed claw pole motor

B. High speed claw pole motor

In our previous work, a high speed claw pole motor as shown in Fig. 2.9 with an SMC stator core was developed [3, 8, 9]. The major motor parameters are tabulated in Table 2.4, and the structure (one pole pitch of one stack) is shown in Fig. 2.10.

Figure 2.11 illustrates the topology of one of the three stacks. The stator consists of the claw poles, the yoke, and the phase winding. The rotor is simply made of a ring PM magnetized in four poles and mounted on the rotor core. The three stator stacks are shifted for 120° (electrical) apart from each other.

Table 2.4 Main dimensions and design parameters

Parameter	Unit	Value
Number of phases	–	3
Rated power	W	2000
Rated frequency	Hz	666.7
Rated speed	rev/min	20,000
Number of poles	–	4
Stator outer diameter	mm	78
Rotor outer diameter	mm	29
Rotor inner diameter	mm	18
Airgap length	mm	1
Axial length	mm	48
Stator core material	–	SOMALOY™ 500
PM material	–	NdFeB

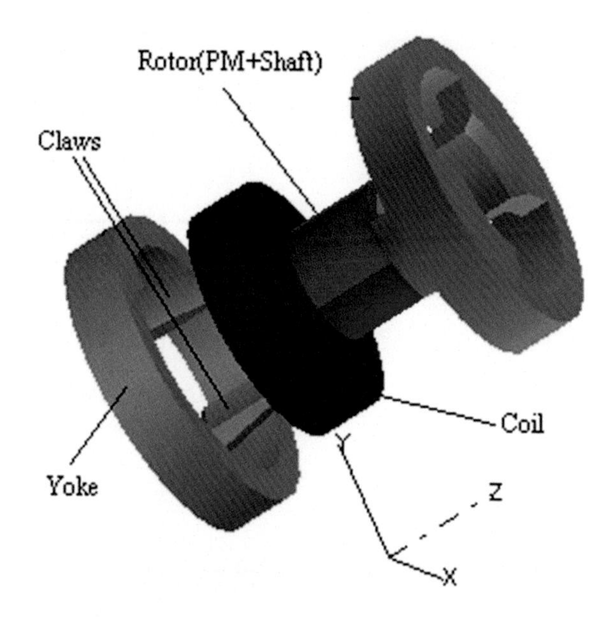

Fig. 2.10 Magnetically relevant parts of one stack of three-phase claw pole motor

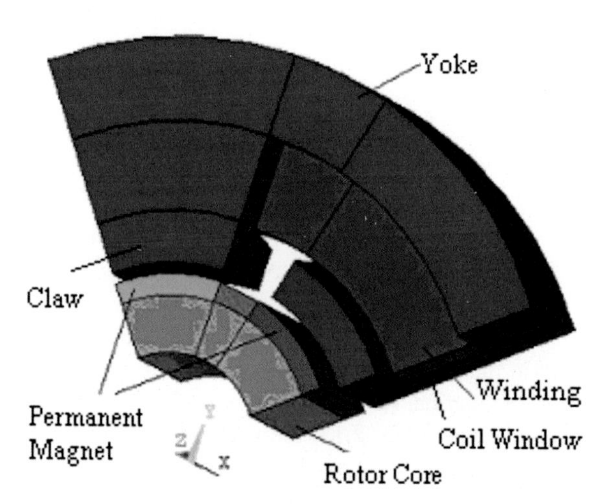

Fig. 2.11 Structure of a high-speed claw pole motor (one pole pitch of one stack)

In general, the geometrical complexity of an electrical machine requires a large thermal network if a high resolution of temperature distribution is required. Instead of using a whole model, the geometrical symmetries of the machine can be used to reduce the size of the model. The distributed thermal properties have been lumped together to form a small thermal network, representing the whole machine. For the calculation of temperature distribution in the SMC motor, a thermal resistance network, as shown in Fig. 2.12, is used. It has ten nodes (the outer air, frame, yoke, and so on). Each node represents a specific part or region of the machine, and the thermal resistances (R_n, $n = 1,\ldots,16$) between the nodes include complex processes,

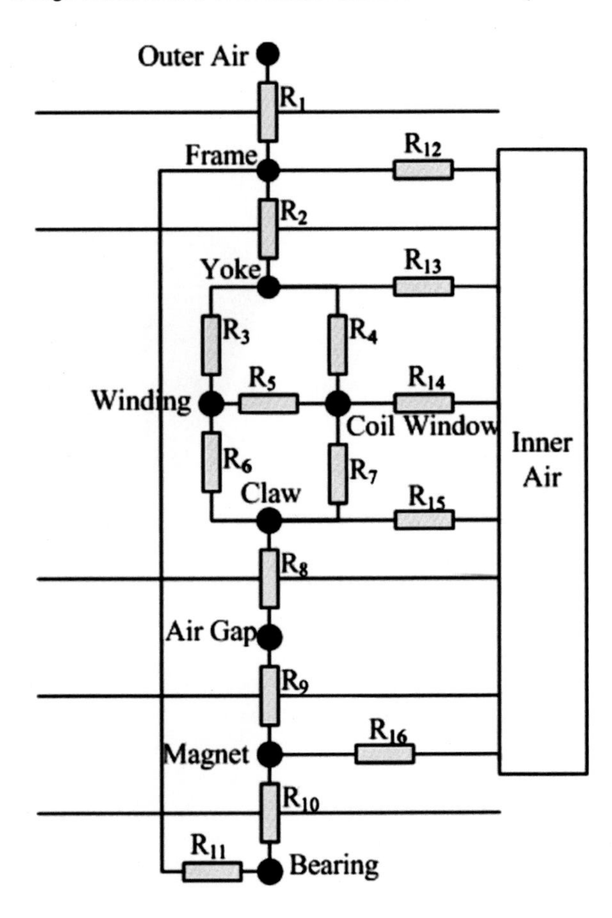

Fig. 2.12 Thermal network of one stack of a high-speed claw pole motor

such as the 2D and 3D heat flow, convection, internal heat generation, and variations in material properties. To account for the three dimensional heat flows at a node, the thermal structure shown in Fig. 2.13 can be employed.

As shown in Fig. 2.13, the thermal resistances of an element are built in three directions, and the heat source if any can be placed at the center point. In this model, the thermal conduction equation can be expressed as

$$\frac{T_b - T_a}{R_{ab}} + \frac{T_c - T_a}{R_{ac}} + \frac{T_d - T_a}{R_{ad}} + \frac{T_e - T_a}{R_{ae}} + \frac{T_f - T_a}{R_{af}} + \frac{T_g - T_a}{R_{ag}} + q_a = C_a \frac{\partial (T_a^{'} - T_a)}{\partial t} \quad (2.16)$$

where T_a, T_b, T_c, T_d, T_e, T_f, and T_g are the temperatures at nodes *a*, *b*, *c*, *d*, *e*, *f*, and *g*, R_{ab}, R_{ac}, R_{ad}, R_{ae}, R_{af}, and R_{ag} the thermal resistance between nodes *a-b*, *a-c*, *a-d*, *a-e*, *a-f*, and *a-g*, respectively, q_a is the heat source, C_a the heat specific, and $T_a^{'}$ the

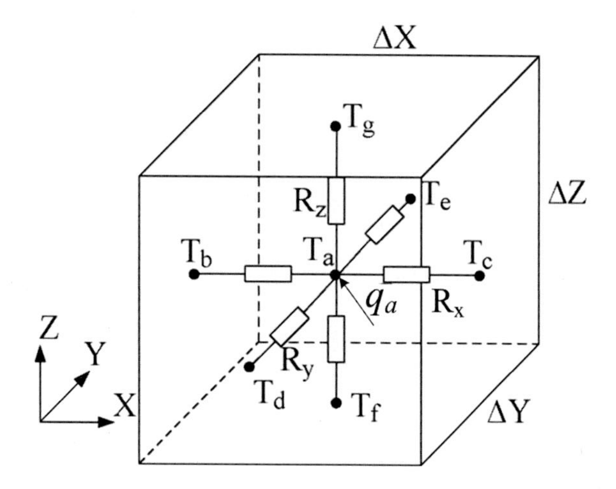

Fig. 2.13 Nodal thermal structure for 3D heat flow

temperature of node a at the next time instant. The thermal resistance in Fig. 2.13 can be calculated by

$$R_{ab} = R_{ac} = \frac{\Delta X}{2\lambda_x \Delta Y \Delta Z} \tag{2.17}$$

$$R_{ad} = R_{ae} = \frac{\Delta Y}{2\lambda_y \Delta X \Delta Z} \tag{2.18}$$

$$R_{af} = R_{ae} = \frac{\Delta Z}{2\lambda_z \Delta X \Delta Y} \tag{2.19}$$

where λ_x, λ_y and λ_z are the thermal conductivities in the x, y and z directions, respectively [8].

The calculation results at no load are 324.6 K in the frame, 326.3 K in the yoke, 330.8 K in the winding, 337.7 K in the claw poles, 334.4 K in the air gap, 331 K in the magnets, and 324.7 K in the bearing.

2.3.3 *Finite Element Model*

In the thermal network, the core loss at each node cannot be obtained easily from the magnetic field calculation. In most cases, the average value is used. Since the core loss distribution is quite different in different positions of the stator core, 3D FEM is used to analyze the temperature distribution in this section. Two design examples investigated in the previous section will be illustrated to show the usage of FEM for the thermal analysis of PM-SMC motors.

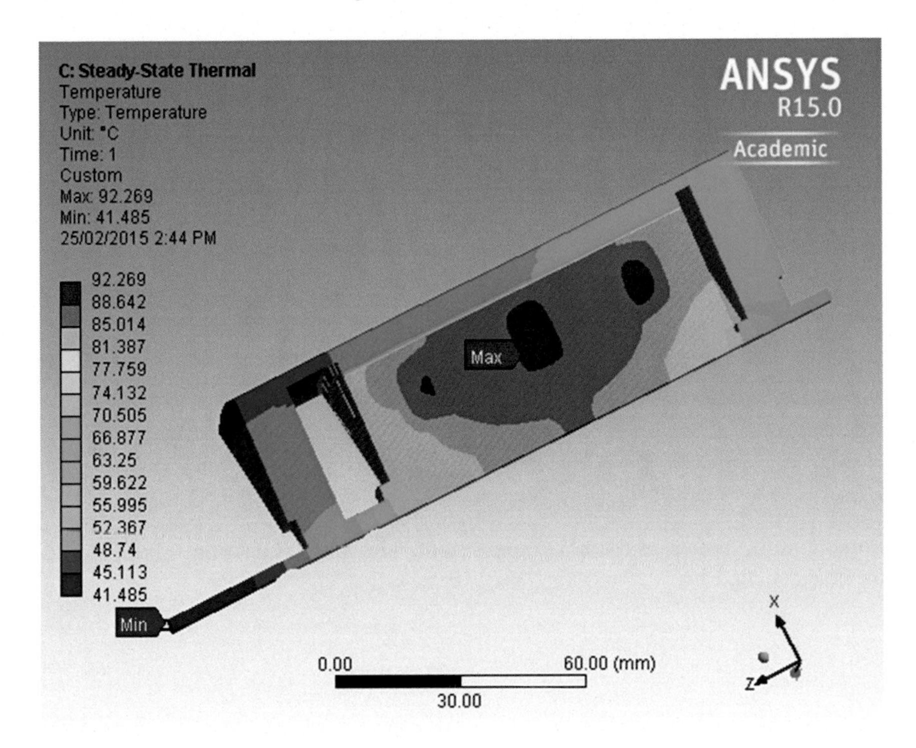

Fig. 2.14 Temperature distribution in the PM-SMC TFM obtained by 3D FEM

Figure 2.14 illustrates the temperature distribution of the PM-SMC TFM based on FEM. As shown, the average temperature rises in the winding is 62.5 °C, which is close to the measured value 65 °C.

Figure 2.15 depicts the distributions of core loss and temperature at full load in the SMC core of the high speed claw pole motor. The temperature is measured by an infrared temperature probe. At 20,000 rev/min and no load, the frame temperature is 331.4 K and the stator yoke temperature is 333.5 K, respectively. The

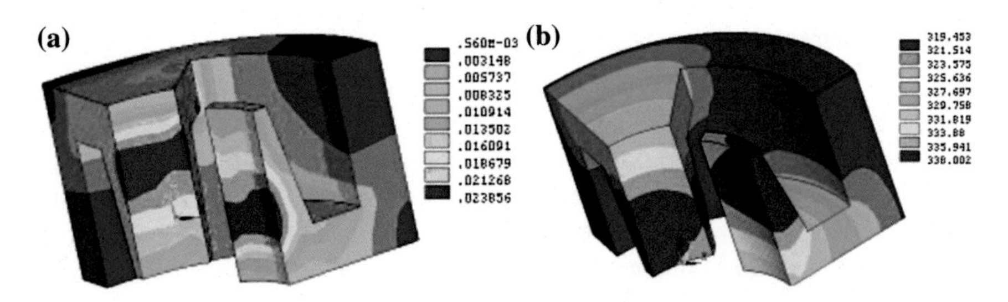

Fig. 2.15 **a** Distributions of core loss, and **b** temperature in SMC core of the high speed claw pole motor

measured temperatures are slightly higher than the FEM results, because the actual loss is greater than the calculation. The FEM method is more accurate than the thermal network method because there are only ten nodes in the network. The advantage of the thermal network is the calculation speed, which is much faster than the FEM method [8, 9].

2.4 Mechanical Design

Mechanical design is another important issue in the design analysis of electrical machines, especially for high speed motors. Generally, the following three aspects should be investigated for the mechanical design analysis:

(1) Mechanical structures and materials,
(2) Field of stress and material strength (including elastic and plastic deformations), and
(3) Modal analysis for vibration and noise.

The first and the second aspects are often noncritical and can be readily satisfied through empirical design, whereas the third one requires special attention for most situations, especially those operated at high frequencies. The modal analysis is generally used to calculate the resonance frequency of the motor in operation. Enough distance between this frequency and the electromagnetic frequency should be designed for motors to avoid resonance. The modal analysis is generally conducted by using the FEM. The two design examples used in the previous section will be employed as follows.

The PM-SMC TFM is operated at 300 Hz, which is relatively high compared with the conventional motors operated at the power frequency of either 50 or 60 Hz. From experience, we only need to do the first-order modal analysis and compare the frequency with the electromagnetic frequency [20]. Figure 2.16 illustrates the first-order modal analysis for this motor of the electromagnetic optimized design. It can be seen that the resonance frequency of the optimal motor is about 4,435 Hz, which is much higher than the electromagnetic frequency of 300 Hz.

Regarding the high speed claw pole motor, because it is operated at high speed, it is essential to carry out a modal analysis to find and adjust the resonant points, so that in practical operation, these frequencies can be avoided. Figure 2.17 shows the vibration patterns and the corresponding resonant frequencies of the rotor structure. These frequencies are well above the operating frequency and therefore have almost no influence to the practical operation. Through the analysis and adjustment, it was found that the bearing stiffness, the shaft length, the shaft diameter and the position of bearing have significant influence on the rotor natural frequency [3].

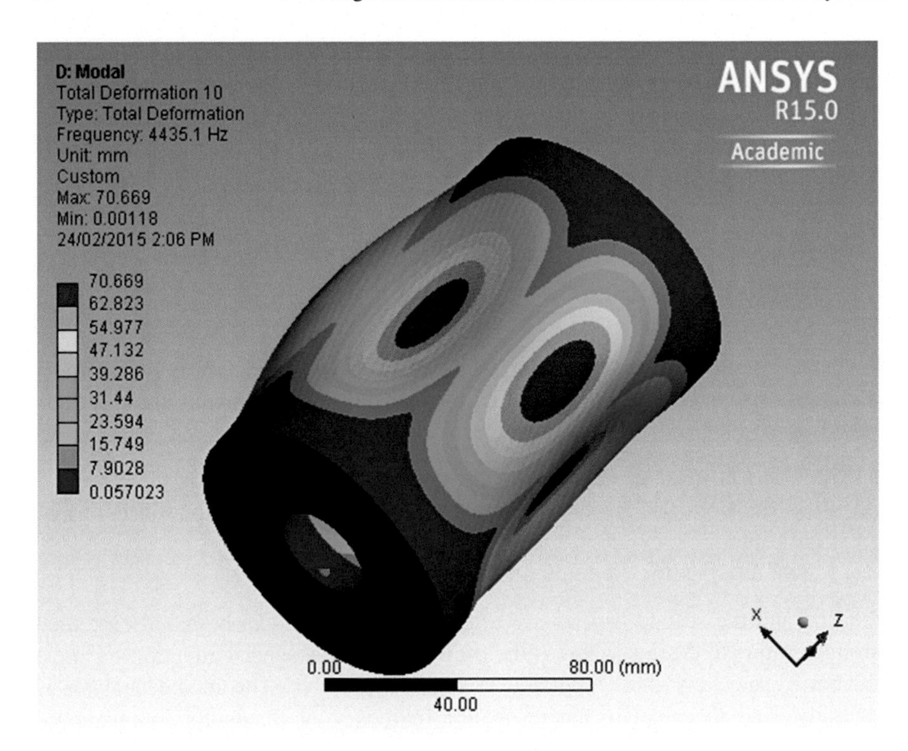

Fig. 2.16 Illustration of first order modal analysis for PM-SMC TFM

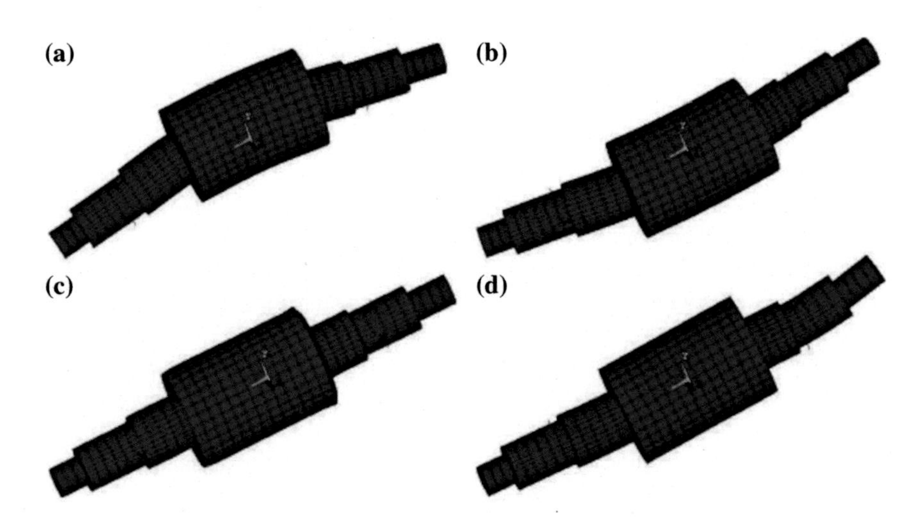

Fig. 2.17 Vibration patterns at **a** 4,102 Hz (Y axis), **b** 4,102 Hz (X axis), **c** 9,562 Hz (Z axis), and **d** 10,321 Hz (Y axis)

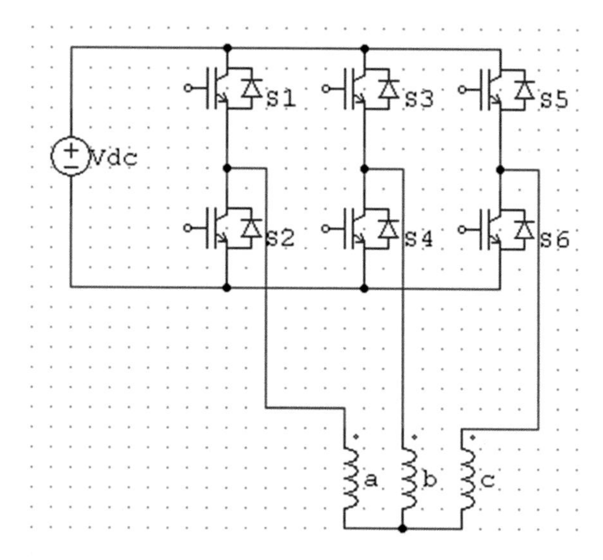

Fig. 2.18 A three-phase inverter

2.5 Power Electronics Design

The design of power electronics for electrical machines and drive systems is also an important and complex stage. Among many aspects in power electronics, the converter/inverter and switching scheme are two main concerns in the design of electrical machines and drive systems.

The converter/inverter is an important component to drive an electrical machine. An inverter, for example, is an electronic apparatus that can convert a DC voltage to an AC voltage of specified waveform, frequency, magnitude, and phase angle. Among many different topologies, the three phase bridge power circuit as shown in Fig. 2.18 has become favorite and standard for use in the control systems of electrical machines. Many different topologies can be obtained from this structure for different applications. For example, two extra switches can be added to establish two bridges for the fault tolerant control scheme [33, 34].

For controlling the waveform, frequency, magnitude, and phase angle of the AC voltage, many switching schemes can be used, such as square wave and sine wave pulse width modulations (PWMs) and space vector modulation (SVM), as well as hard and soft switching.

2.6 Control Algorithms Design

Control algorithms play an important role in the determination of dynamic and state-state performances of electrical drive systems. Various control algorithms have been developed and employed successfully in commercial drive systems, such as the six-step control, FOC, DTC and MPC [35–39].

While FOC is commonly used in various high performance electrical drive system, the merits of DTC are simple in structure (thus low cost), fast dynamic response, and strong robustness against motor parameter variation [40–42]. The major advantages that affect the commercial application of the conventional DTC are large torque and flux ripples, variable switching frequency, and excessive acoustic noises.

To overcome these problems, many methods have been proposed in the literature. One of them is to apply the technique of SVM to DTC, known as the SVM-DTC. In the conventional DTC, the switching table only includes a limited number of voltage vectors with fixed amplitudes and positions. The implementation of SVM enables the generation of an arbitrary voltage vector with any amplitude and position [43–48]. In this way, the SVM-DTC can generate the torque and flux more accurately to eliminate the ripples. Another merit of SVM-DTC is that the sampling frequency required is constant and lower than that of the conventional DTC.

Recently, the MPC has attracted increasing attention in industry and academic communities [49–54]. In the SVM-DTC, the power converter with modulation can be considered as a gain in controller design. In the predictive control methods, the discrete nature of power converters is taken into account by considering the converter and the motor from a systemic viewpoint. There are various different versions of predictive control algorithms, differing in the principle of vector selection, number of the applied vectors and predictive horizon.

The conventional DTC and MPC are similar in that they both select only one voltage vector in each sampling period. This can result in overregulation, leading to large torque and flux ripples and acoustic noise.

As all the design examples used in this book are permanent magnet synchronous machines (PMSMs), several control algorithms will be presented with details for PMSMs in the following sections. Numerical and experimental examples will be presented for some of them.

2.6.1 Six-Step Control

The six-step control method was oriented to drive brushless DC (BLDC) motors with trapezoidal back EMF waveforms. In many applications, however, the trapezoidal excitation is also used to drive PMSMs with sinusoidal back EMF waveforms because the trapezoidal excitation or six-step method based drive is robust and low cost [35].

In the six-step control scheme, the stationary reference frame is always used to model the PMSM. The phase variables are used to express the machine equations as they can account for the real waveforms of the back EMF and phase current. Assuming that the resistances of three phase stator windings are equal, the three phase voltage equations of the motor can be written as

$$\begin{bmatrix} v_a \\ v_b \\ v_c \end{bmatrix} = \begin{bmatrix} R_s & 0 & 0 \\ 0 & R_s & 0 \\ 0 & 0 & R_s \end{bmatrix} \begin{bmatrix} i_a \\ i_b \\ i_c \end{bmatrix} + \frac{d}{dt} \left\{ \begin{bmatrix} L_{aa} & L_{ba} & L_{ca} \\ L_{ba} & L_{bb} & L_{cb} \\ L_{ca} & L_{cb} & L_{cc} \end{bmatrix} \begin{bmatrix} i_a \\ i_b \\ i_c \end{bmatrix} \right\} + \begin{bmatrix} e_a \\ e_b \\ e_c \end{bmatrix} \tag{2.20}$$

where v_a, v_b, and v_c are the phase voltages, i_a, i_b, and i_c the phase currents, e_a, e_b, and e_c the phase back EMF, R_s is the phase resistance, and $\begin{bmatrix} L_{aa} & L_{ba} & L_{ca} \\ L_{ba} & L_{bb} & L_{cb} \\ L_{ca} & L_{cb} & L_{cc} \end{bmatrix}$ the inductance matrix, including both the self-and mutual-inductances.

Assuming further that the reluctance is independent of the rotor position, one can obtain

$$\begin{cases} L_a = L_b = L_c = L_s \\ L_{ab} = L_{ca} = L_{bc} = M \end{cases} \tag{2.21}$$

As $i_a + i_b + i_c = 0$ for a symmetric three phase system, the voltage equation can be simplified as

$$\begin{bmatrix} v_a \\ v_b \\ v_c \end{bmatrix} = \begin{bmatrix} R_s & 0 & 0 \\ 0 & R_s & 0 \\ 0 & 0 & R_s \end{bmatrix} \begin{bmatrix} i_a \\ i_b \\ i_c \end{bmatrix} + \frac{d}{dt} \left\{ \begin{bmatrix} L-M & 0 & 0 \\ 0 & L-M & 0 \\ 0 & 0 & L-M \end{bmatrix} \begin{bmatrix} i_a \\ i_b \\ i_c \end{bmatrix} \right\} + \begin{bmatrix} e_a \\ e_b \\ e_c \end{bmatrix} \tag{2.22}$$

Assuming linear system, the machine model in state space form can be expressed as

$$\frac{d}{dt} \begin{bmatrix} i_a \\ i_b \\ i_c \end{bmatrix} = \begin{bmatrix} 1/(L-M) & 0 & 0 \\ 0 & 1/(L-M) & 0 \\ 0 & 0 & 1/(L-M) \end{bmatrix} \left\{ \begin{bmatrix} v_a \\ v_b \\ v_c \end{bmatrix} - \begin{bmatrix} R_s & 0 & 0 \\ 0 & R_s & 0 \\ 0 & 0 & R_s \end{bmatrix} \begin{bmatrix} i_a \\ i_b \\ i_c \end{bmatrix} - \begin{bmatrix} e_a \\ e_b \\ e_c \end{bmatrix} \right\} \tag{2.23}$$

The generated electromagnetic torque is given by

$$T_e = (e_a i_a + e_b i_b + e_c i_c)/\omega_m \tag{2.24}$$

where ω_m is the mechanical angular speed of the rotor.

The mechanical equation of the machine is

$$T_e = \frac{d\omega_m}{dt} J + F\omega_m + T_L \tag{2.25}$$

where J is the inertia of the machine rotating parts, F the friction coefficient, and T_L the load torque on the rotor shaft.

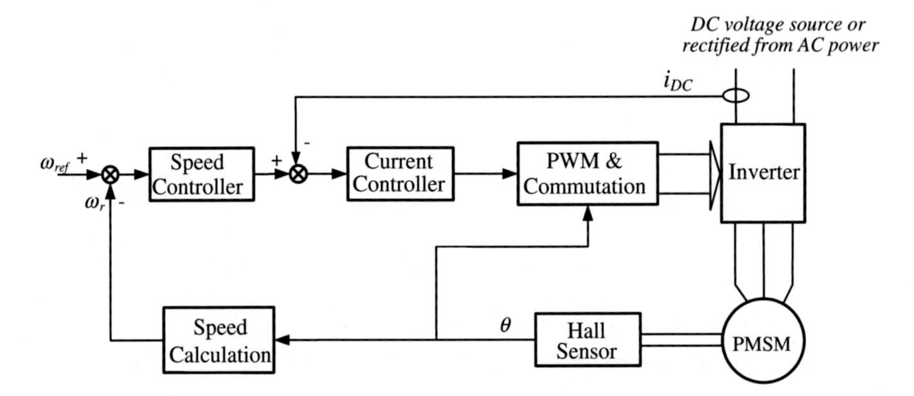

Fig. 2.19 Block diagram of PMSM six-step drive system

Figure 2.19 shows the block diagram of six-step drive scheme. The drive system is operated with the feedback information of rotor position, which is obtained at fixed points, typically every 60 electrical degrees for commutation of the phase currents.

The 120° conduction mode is applied to drive the PMSM. The voltage may be applied to the motor every 120° (electrical), with a current limit to hold the phase currents within the motor's capabilities. Because the phase currents are excited in synchronism with the back EMF, a constant torque is generated. A simulation model is built in MATLAB/SIMULINK as shown in Fig. 2.20.

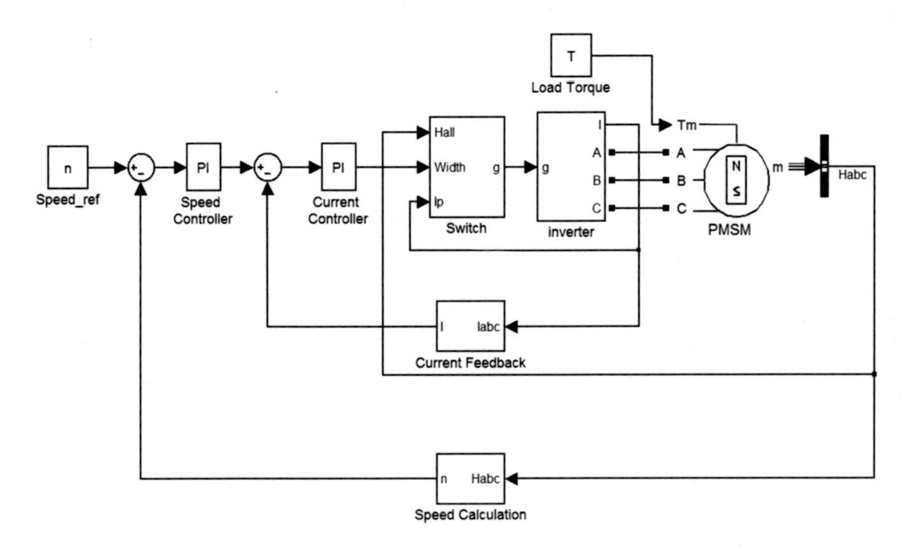

Fig. 2.20 Simulation block diagram of six-step controlled PMSM drive system

As shown in Fig. 2.20, the rotor position information comes from the Hall effect sensors, which are integrated in the machine model in MATLAB/SIMULINK. The resolution of the feedback signals is only 60° (electrical). Since most applications require a stable speed, a speed feedback loop is employed. The rotor speed information can be deduced from the low resolution Hall signals, which is marked as Speed Calculation in Fig. 2.20. Typically, the average speed in one 60° section is used as the speed feedback.

However, by using the average speed, there is always a lag when the motor speed is not constant in accelerating or other dynamic state. To overcome this, the rotor position can be expressed in Taylor's series as the following:

$$\theta(t) = \theta_k(t) + \theta_{1k}^{(1)}(t - t_k) + \frac{\theta_{2k}^{(2)}}{2!}(t - t_k)^2 + \cdots \quad (2.26)$$

where t_k is the last commutation time, $\theta_{1k}^{(1)} = \frac{\pi/3}{t_k - t_{k-1}}$ the average speed of last section, and $\theta_{2k}^{(2)} = \frac{\theta_{1k}^{(1)} - \theta_{1(k-1)}^{(1)}}{t_k - t_{k-1}}$ the average acceleration of last section.

As shown above, with the higher order calculation, more accurate speed and position information can be deduced, whereas the computing cost rises. As a compromise, in some situations, the following equations are used to estimate the rotor position and speed:

$$\begin{cases} \theta(t) = \theta_k(t) + \theta_{1k}^{(1)}(t - t_k) + \frac{\theta_{2k}^{(2)}}{2!}(t - t_k)^2 \\ \omega(t) = \theta_{1k}^{(1)}(t) + \theta_{2k}^{(2)}(t - t_k) \end{cases} \quad (2.27)$$

2.6.2 *Field Oriented Control*

For a PMSM under sinusoidal excitations, the original voltage equations can be expressed in the stationary reference frame as the following

$$\begin{bmatrix} v_a \\ v_b \\ v_c \end{bmatrix} = R_s \begin{bmatrix} i_a \\ i_b \\ i_c \end{bmatrix} + \frac{d}{dt} \begin{bmatrix} \lambda_a \\ \lambda_b \\ \lambda_c \end{bmatrix} \quad (2.28)$$

where λ_a, λ_b, and λ_c are the flux linkages of phases *a*, *b*, and *c*, respectively.

Equation (2.28) represents a system of differential equations with time varying (periodic) coefficients. For sinusoidally distributed windings, a Park-Clark transformation can be used to transform the above equations to a system of differential equations with constant coefficients, represented in a *d-q* coordinate frame attached to the rotor. The reference frames are shown in Fig. 2.21.

The Park-Clark orthogonal transformation can be expressed in the matrix form as

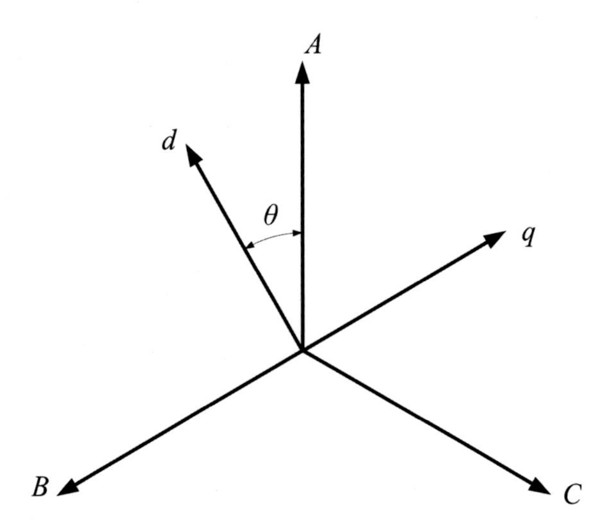

Fig. 2.21 Stationary and rotating reference frames

$$\begin{bmatrix} \sigma_d \\ \sigma_q \\ \sigma_0 \end{bmatrix} = \sqrt{\frac{2}{3}} \begin{bmatrix} \cos\theta & \cos\left(\theta - \frac{2}{3}\pi\right) & \cos\left(\theta + \frac{2}{3}\pi\right) \\ -\sin\theta & -\sin\left(\theta - \frac{2}{3}\pi\right) & -\sin\left(\theta + \frac{2}{3}\pi\right) \\ \frac{1}{\sqrt{2}} & \frac{1}{\sqrt{2}} & \frac{1}{\sqrt{2}} \end{bmatrix} \begin{bmatrix} \sigma_a \\ \sigma_b \\ \sigma_c \end{bmatrix} \tag{2.29}$$

where θ is defined as the angle between two reference frames.

The subscripts d, q, and 0 in (2.29) represent some fictitious windings attached to the rotor. The variables σ_d, σ_q, σ_0, σ_a, σ_b, and σ_c may represent voltages, currents, or flux linkages. As a result, the transformed set of electrical equations describing the behavior of PMSM in the d-q rotating frame become

$$\begin{cases} v_d = R_s i_d + \frac{d}{dt}\lambda_d - \lambda_q \frac{d\theta}{dt} \\ v_q = R_s i_q + \frac{d}{dt}\lambda_q + \lambda_d \frac{d\theta}{dt} \\ v_0 = R_s i_0 + \frac{d}{dt}\lambda_0 \end{cases} \tag{2.30}$$

where v_d, v_q, and v_0 are the phase voltages, i_d, i_q, and i_0 the phase currents, and λ_d, λ_q, and λ_0 the phase flux linkages.

For the linear PMSM model, the magnetic saturation saliency is not considered. The flux linkages of the d- and q-axes can be further expressed as

$$\begin{cases} \lambda_d = L_d i_d + \lambda_m \\ \lambda_q = L_q i_q \end{cases} \tag{2.31}$$

where L_d and L_q are the constant d- and q-axes inductances, respectively, and λ_m is the flux linkage generated by the rotor PMs.

On the other hand, the voltage equation of the 0 axis in (2.30) is usually ignored by assuming well-balanced three-phase windings for the controller design.

Therefore, the electrical voltage equations in the rotor reference frame can be rewritten as

$$\begin{cases} v_d = R_s i_d + L_d \frac{di_d}{dt} - L_q i_q \frac{d\theta}{dt} \\ v_q = R_s i_q + L_q \frac{di_q}{dt} + (L_d i_d + \lambda_m) \frac{d\theta}{dt} \end{cases} \tag{2.32}$$

The torque expression after the application of the transformation becomes

$$\begin{aligned} T_e &= \frac{3}{2} p \left(\lambda_d i_q - \lambda_q i_d \right) \\ &= \frac{3}{2} p \left[\lambda_m i_q + \left(L_d - L_q \right) i_d i_q \right] \end{aligned} \tag{2.33}$$

where p is the number of pole pairs.

By this transformation, the flux and torque control of the PMSM are decoupled. The q-axis current, in the FOC method, is regulated to produce sufficient torque while the d-axis current is controlled to modify the air-gap flux linkage. For normal operation, the d-axis current is set to zero to achieve the maximum torque-to-ampere ratio, and for the flux weakening control, the d-axis current is modified to weaken the air-gap flux.

The reference speed value is the main input for the drive system, and the electromagnetic torque and rotor speed are the output. Two feedback loops, current or torque loop and speed loop, are added to provide desired performance. The output of the speed controller will be the reference value for the q-axis current while the d-axis current is set to zero. Both of the d- and q-axes currents are controlled to generate the torque and achieve the maximum efficiency drive. Figure 2.22 shows the implementation diagram of the typical FOC scheme, where the traditional PWM method is applied for the variable speed drive by the vectorial variable voltage and variable frequency control strategy.

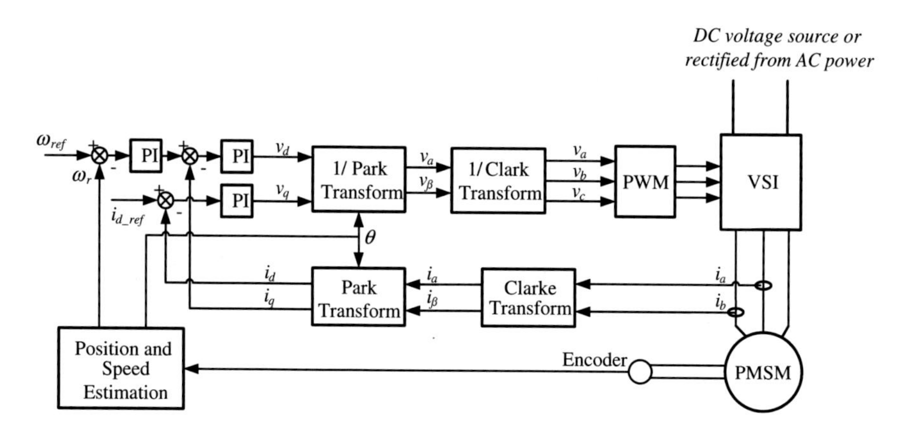

Fig. 2.22 Block diagram of FOC scheme for PMSM drive

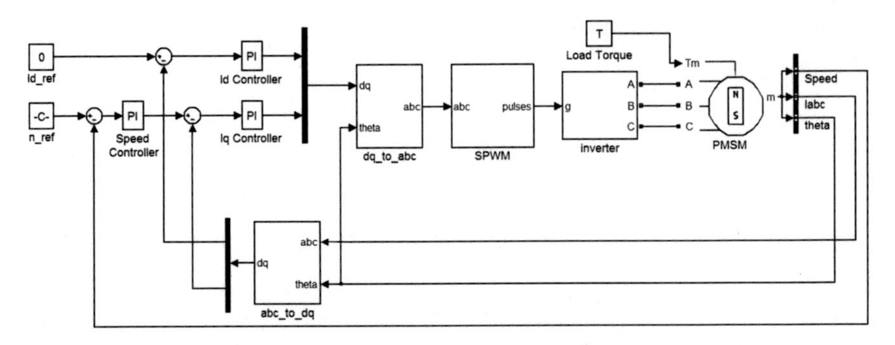

Fig. 2.23 Simulation block diagram of typical FOC based PMSM drive system

Similar to the six-step method, a simulation model of the FOC scheme based PMSM drive is built in MATLAB/SIMULINK. The sinusoidal back EMF machine model is selected from the SimPowerSystem tool box, in which the current sensors and rotor position sensor are integrated. The Park and Clark transformations are synthesized as one '*abc_to_dq*' block to transfer the variables between the stationary and rotating reference frames, as shown in Fig. 2.23. Two discrete PI controllers are used for the speed and current feedback loops.

The traditional triangulation PWM generation technique is applied. A triangular carrier wave sampling signal is compared directly with a sinusoidal modulating wave to determine the switching instants, and therefore the resultant pulse widths.

2.6.3 Direct Torque Control

In the DTC strategy, the flux linkage and torque are calculated in the two-phase stator reference frame, i.e., the α-β frame, which is transformed from the three-phase a-b-c reference frame by using the Clark transformation. The Clark transformation can be expressed in the matrix form as

$$\begin{bmatrix} \sigma_\alpha \\ \sigma_\beta \end{bmatrix} = \sqrt{\frac{2}{3}} \begin{bmatrix} 1 & -\frac{1}{2} & -\frac{1}{2} \\ 0 & \frac{\sqrt{3}}{2} & -\frac{\sqrt{3}}{2} \end{bmatrix} \begin{bmatrix} \sigma_a \\ \sigma_b \\ \sigma_c \end{bmatrix} \tag{2.34}$$

After the measured phase voltages and currents are transformed to the α-β frame, the flux linkage components of the α- and β-axes can be calculated as

$$\begin{cases} \lambda_\alpha = \int (v_\alpha - R_s i_\alpha) dt \\ \lambda_\beta = \int \left(v_\beta - R_s i_\beta\right) dt \end{cases} \tag{2.35}$$

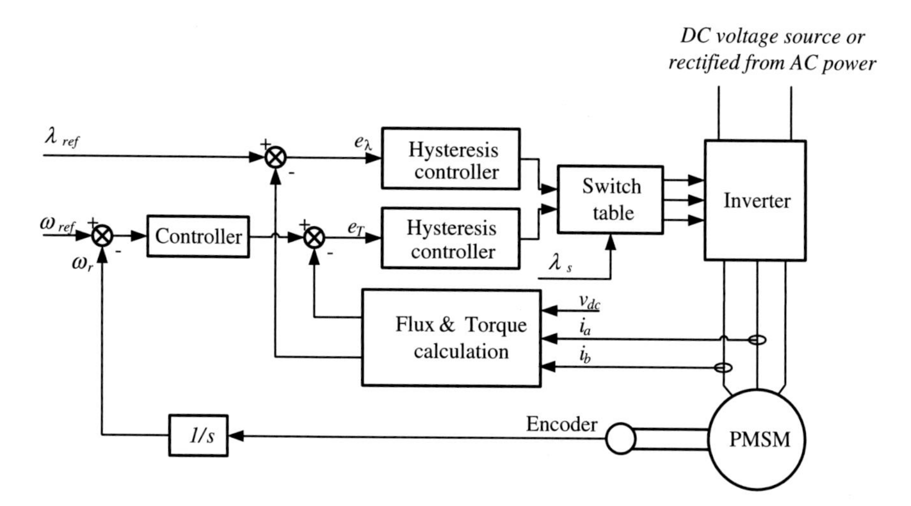

Fig. 2.24 Block diagram of typical DTC scheme based PMSM drive

The torque observer can then be designed as

$$T_e = \frac{3}{2} \cdot \frac{p_m}{2} \left(\lambda_\alpha i_\beta - \lambda_\beta i_\alpha\right) \tag{2.36}$$

Figure 2.24 shows the block diagram of a typical DTC scheme for PMSM drive. Two hysteresis controllers are applied to the flux linkage and torque control loops. The calculated flux linkage is also sent to the switching table to identify the current flux vector position.

From (2.35), the stator flux linkage is

$$\lambda_s = \int (v_s - R_s i_s) dt \tag{2.37}$$

where v_s and i_s are the stator voltage and current spatial vectors, respectively.

In the case of a PMSM, λ_s always varies even when the zero voltage vectors are applied because of the rotating rotor magnets, and thus, zero voltage vectors are not used for DTC driven PMSM. λ_s should always be in motion with respect to the rotor flux.

According to (2.36), the electromagnetic torque can be controlled effectively by controlling the amplitude and rotating speed of λ_s. For counter-clockwise operation, if the actual torque is smaller than the reference, the voltage vectors that keep λ_s rotating in the same direction are selected. The angle increases as fast as it can, and the actual torque increases as well. Once the actual torque is greater than the reference, the voltage vectors that keep λ_s rotating in the reverse direction are selected instead of the zero voltage vectors. The angle decreases, so does the torque. By selecting the voltage vectors in this way, λ_s will rotate all the time in the

Table 2.5 Switching table of typical DTC scheme for PMSM drive

Δe_λ	Δe_T	θ					
		θ_1	θ_2	θ_3	θ_4	θ_5	θ_6
1	1	$V_2(110)$	$V_3(010)$	$V_4(011)$	$V_5(001)$	$V_6(101)$	$V_1(100)$
	0	$V_6(101)$	$V_1(100)$	$V_2(110)$	$V_3(010)$	$V_4(011)$	$V_5(001)$
0	1	$V_3(010)$	$V_4(011)$	$V_5(001)$	$V_6(101)$	$V_1(100)$	$V_2(110)$
	0	$V_5(001)$	$V_6(101)$	$V_1(100)$	$V_2(110)$	$V_3(010)$	$V_4(011)$

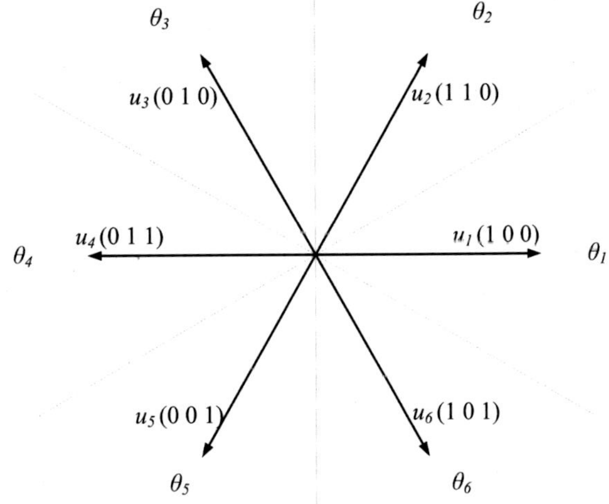

Fig. 2.25 Voltage vectors and spatial sector definition

direction determined by the output of the hysteresis controller for the torque. The switching table for controlling both the amplitude and rotating direction is shown in Table 2.5, in which the inverter voltage vector and spatial sector definitions are illustrated in Fig. 2.25.

Figure 2.26 shows the simulation model built based on the typical DTC scheme. The inverter switching status and DC bus voltage are utilized to calculate the stator voltage. The stator flux linkage is obtained in the observer. The traditional two-level hysteresis controllers are applied and the switching table is designed based on Table 2.5.

2.6.4 Model Predictive Control

The principle of MPC was introduced for industrial control applications in the 1970s after the publication of this strategy in the 1960s. The MPC requires great computational effort and it has been formerly limited to slowly varying systems, such as chemical processes. With the availability of inexpensive high computing

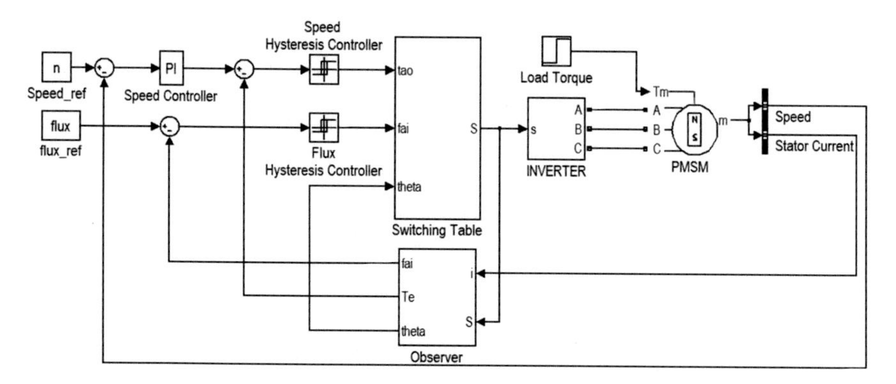

Fig. 2.26 Simulation block diagram of typical DTC based PMSM drive system

power microcomputers and modern digital control techniques, MPC is able to be applied to electrical drive systems [36, 55, 56].

Different from the employment of hysteresis comparators and the switching table in conventional DTC, the principle of vector selection in MPC is based on evaluating a defined cost function. The selected voltage vector from the conventional switching table in DTC may not necessarily be the best one for the purposes of torque and flux ripple reduction. Since there are limited discrete voltage vectors in the two-level inverter-fed PMSM drives, it is possible to evaluate the effects of each voltage vector and select the one minimizing the cost function.

The key technology of MPC lies in the definition of the cost function, which is related to the control objectives. The greatest concerns of PMSM drive applications are the torque and stator flux, and thus, the cost function is defined in such a way that both the torque and stator flux at the end of control period are as close as possible to the reference values. In this book, the cost function is defined as

$$\begin{aligned} \text{min:} \quad & G = |T_e^* - T_e^{k+1}| + k_1 \big| |\psi_s^*| - |\psi_s^{k+1}| \big| \\ & \text{s.t. } u_s^k \in \{V_0, V_1, \ldots, V_7\} \end{aligned} \tag{2.38}$$

where T_e^* and ψ_s^* are the reference torque and flux, T_e^{k+1} and ψ_s^{k+1} the predicted values of torque and flux, respectively, and k_1 is the weighting factor. Because the physical natures of electromagnetic torque and stator flux are different, the weighting factor k_1 is introduced to unify these terms. In this work, k_1 is selected to be T_n/ψ_n, where T_n and ψ_n are the rated values of torque and stator flux, respectively. It should be noted that when a null vector is selected, the specific state (V_0 or V_7) will be determined based on the principle of minimal switching commutations, which is related to the switching states of the previous voltage vector.

The voltage equations in the *d-q* reference frame are as follows:

$$u_d = R_s i_d + L_d \frac{di_d}{dt} - \omega L_q i_q \tag{2.39}$$

$$u_q = R_s i_q + L_q \frac{di_q}{dt} + \omega L_d i_d + \omega \psi_f \tag{2.40}$$

Given the voltage and current values at sampling instant k, the predicted current, torque and flux at instant $k + 1$ can be expressed as follows:

$$i_d^{k+1} = i_d^k + \frac{1}{L_d}\left(-R_s i_d^k + \omega^k L_q i_q^k + u_d^k\right) T_s \tag{2.41}$$

$$i_q^{k+1} = i_q^k + \frac{1}{L_q}\left(-\omega^k L_d i_d^k - R_s i_q^k + u_q^k - \omega^k \psi_f\right) T_s \tag{2.42}$$

$$\psi_s^{k+1} = \left(L_d i_d^{k+1} + \psi_f\right) + jL_q i_q^{k+1} \tag{2.43}$$

$$T_e^{k+1} = \frac{3}{2} p \psi_s^{k+1} i_s^{k+1} \tag{2.44}$$

where i_d^{k+1} and i_q^{k+1} are the predicted values of stator current for the sampling instant $k + 1$, T_s is the sampling period, T_e^{k+1} and ψ_s^{k+1} are the predicted values of torque and flux, respectively, which are also the main concerns for the cost function in the following MPC control scheme [1, 36, 49].

The block diagram of MPC is shown in Fig. 2.27. The inputs of the system are the reference and estimated values of torque and flux. By evaluating the effects of each voltage vector when applied to the machine, the voltage vector which minimizes the difference between the reference and predicted values is first selected, and then it is generated by the inverter.

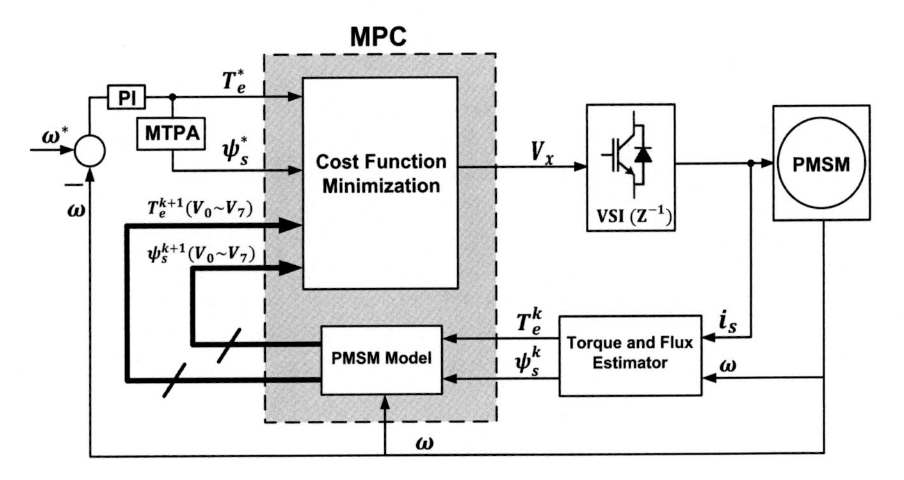

Fig. 2.27 Block diagram of MPC drive system in MATLAB/SIMULINK

2.6.4.1 One-Step Delay Compensation

The cost function in (2.38) assumes that all calculations and judgments are implemented at the kth instant and the selected vector will be applied immediately. However, in practical digital implementation, this assumption is not true and the applied voltage vector is not applied until the (k + 1)th instant.

In other words, for the duration between the kth and (k + 1)th instants, the applied rotor voltage vector u_s^k has been decided by the value in the (k-1)th instant and the evolutions of ψ_s and T_e for this duration are uncontrollable. What is left to be decided is actually the stator voltage vector u_s^{k+1}, which is applied at the beginning of the (k + 1)th instant. To eliminate this one step delay, the variables of ψ_s^{k+2} and T_e^{k+2} should be used rather than ψ_s^{k+1} and T_e^{k+1} for the evaluation of the cost function in (2.38). This fact is clearly illustrated in Fig. 2.28, where x indicates the state variables of a dynamic system and u is the input to be decided. For PMSM, x represents torque or stator flux value.

To eliminate the one-step delay in digital implementation, the cost function in (2.38) should be changed to (2.45) as shown below

$$\begin{aligned} \text{min:} \quad & G = |T_e^* - T_e^{k+2}| + k_1 \big| |\psi_s^*| - |\psi_s^{k+2}| \big| \\ & \text{s.t. } u_s^k \in \{V_0, V_1, \ldots, V_7\} \end{aligned} \tag{2.45}$$

Obtaining ψ_s^{k+2} and T_e^{k+2} in (2.45) requires a two-step prediction. To obtain the best voltage vector minimizing the cost function in (2.45), each possible configuration for u_e^{k+1} will be evaluated to obtain the value at the (k + 2)th instant.

2.6.4.2 Linear Multiple Horizon Prediction

A linear multiple horizon prediction formula is introduced in this section. This formula incorporates two formulas. The first one is the same as in (2.38). The linear

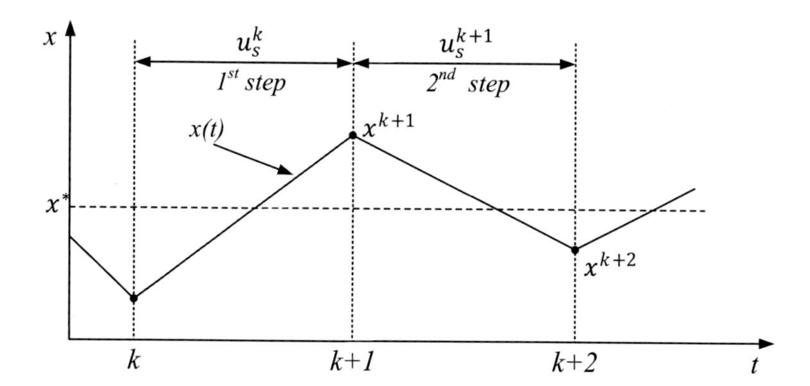

Fig. 2.28 One-step delay in digital control systems

multiple horizon prediction formula, which is multiplied by a factor A, considers the errors in the (k + N)th instant (N > 1). Different from the model-based predictions for ψ_s^{k+1} and T_e^{k+1}, the stator flux and torque at the (k + N)th instant are predicted from the value at the kth and (k + 1)th instants using linear extrapolations, which are expressed as

$$T_e^{k+N} = T_e^k + (N-1)(T_e^{k+1} - T_e^k) \tag{2.46}$$

$$|\psi_s^{k+N}| = \left||\psi_s^k| + (N-1)\left||\psi_s^{k+1}| - |\psi_s^k|\right|\right| \tag{2.47}$$

The expression of the proposed cost function is

$$\begin{aligned} \text{min:} \quad G = {} & |T_e^* - T_e^{k+1}| + k_1\left||\psi_s^*| - |\psi_s^{k+1}|\right| \\ & + A\left(|T_e^* - T_e^{k+N}| + k_1\left||\psi_s^*| - |\psi_s^{k+N}|\right|\right) \quad s.t.\, u_s^k \in \{V_0, V_1, \ldots, V_7\} \end{aligned} \tag{2.48}$$

2.6.5 *Numerical and Experimental Comparisons of DTC and MPC*

2.6.5.1 Numerical Simulation

In this section, the simulation tests of DTC and MPC are carried out by using Matlab/Simulink. The parameters of the motor are listed in Table 2.6. The sampling frequency of both methods is set to 5 kHz. The values of control parameters are $k_1 = 25.4$, $A = 0.1$, and $N = 10$ [36].

This simulation test combines start-up, steady-state and external load tests. The motor starts up from 0 s with several reference speeds (500 rev/min, 1000 rev/min, 1500 rev/min and 2000 rev/min). After reaching the reference speed, the motor maintains the speed for at least 0.2 s and an external load is applied at 0.3 s. Figures 2.29, 2.30, 2.31 and 2.32 show the combined load test for four control strategies for one reference speed, 1000 rev/min. From top to bottom, the curves are the stator current, stator flux, torque, motor speed, and switching frequency, respectively. The test results for other speed situations can be found in [36].

Table 2.6 Motor parameters

Number of pole pairs	p	3
Permanent magnet flux	ψ_f	0.1057 Wb
Stator resistance	R_s	1.8 Ω
d- and q-axis inductance	L_d, L_q	15 mH
Rated torque	T_N	4.5 Nm
DC bus voltage	V_{dc}	200 V

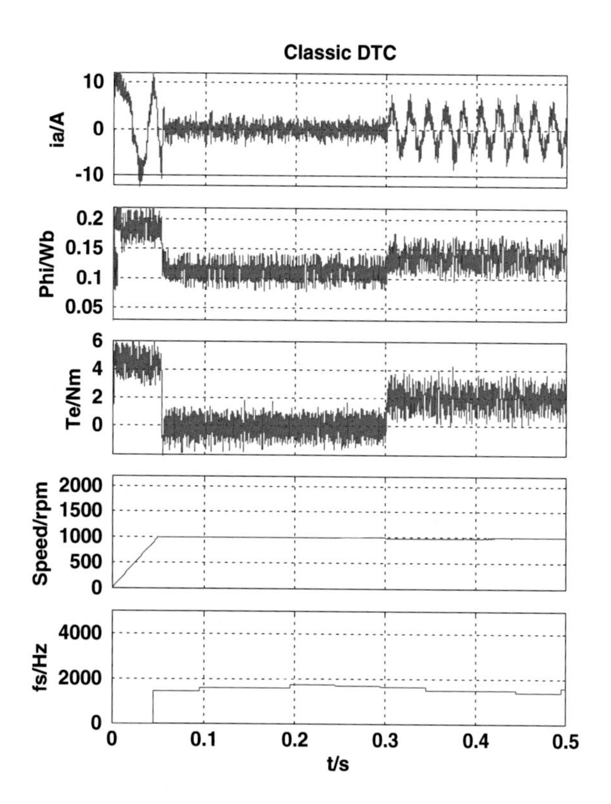

Fig. 2.29 Combined load test for DTC

By comparing Fig. 2.30 with Fig. 2.29, it is shown that the torque and flux ripples of MPC are lower than that of DTC. In Fig. 2.31, MPC with one-step delay compensation (indicated as MPC + comp) presents torque and flux ripples even lower than MPC along with an increase in switching frequency. Figure 2.32 illustrates the responses by using cost function (2.48), where factor A is included in the simulation. As shown, the introduction of linear multiple horizon prediction (factor A, and indicated as MPC + A) can greatly reduce the switching frequency only with a quite limited degradation of torque and flux ripples. As shown, all these methods present similar dynamic performance and the motor can reach the reference speed rapidly. When the load was applied, the motor speed returned to its original value in a very short time period.

The recorded data from 0.1 to 0.3 s are picked to calculate the torque and flux ripples (obtained by standard deviations). The torque and flux ripples of these control methods are summarized in Table 2.7. A segment (three periods) of the stator current of phase A is used to calculate the total harmonic distortion (THD) and current harmonic spectrum.

As shown, MPC can achieve lower torque ripple than that of DTC as proven. However, MPC's characteristic in flux ripple reduction is quite unstable. With the help of one-step delay compensation, the steady-state performance of MPC is

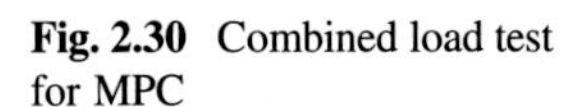
Fig. 2.30 Combined load test for MPC

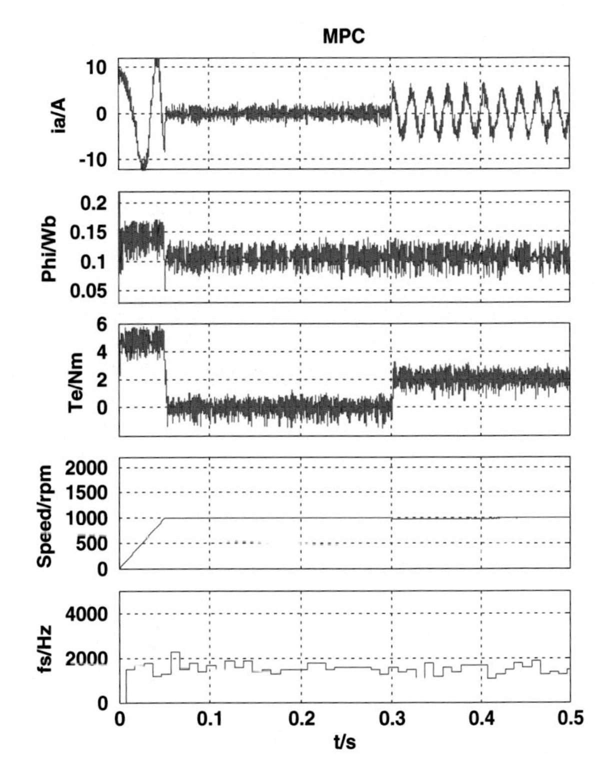

improved significantly. It should be noticed that the switching frequency has increased by almost two times in most tests when one-step delay is compensated. The introduction of linear multiple horizon prediction can effectively reduce the switching frequency and flux ripple. However, its ability on torque ripple reduction is quite insignificant.

2.6.5.2 Experimental Testing

In addition to the simulation study, the control methods mentioned above are further experimentally tested on a two-level inverter-fed PMSM motor drive. The experimental setup is illustrated in Fig. 2.33. A dSPACE DS1104 PPC/DSP control board is employed to implement the real-time algorithm coding using C language. A three phase intelligent power module equipped with an insulated-gate bipolar transistor (IGBT) is used as an inverter. The gating pulses are generated in the DS1104 board and then sent to the inverter. The load is applied using a programmable dynamo-meter controller DSP6000 (Fig. 2.34). A 2500-pulse incremental encoder is equipped to obtain the rotor speed of PMSM. All experimental results are recorded by the ControlDesk interfaced with DS1104 and PC at 5 kHz sampling frequency [36].

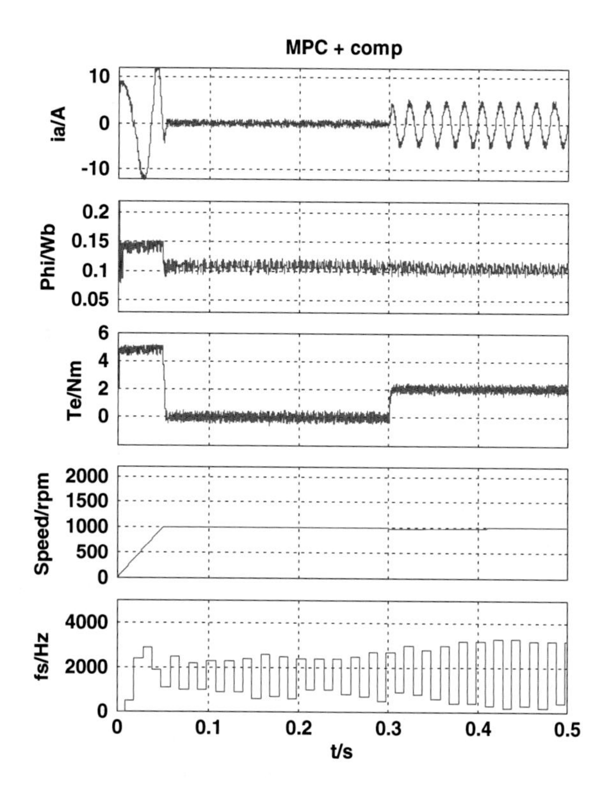

Fig. 2.31 Combined load test for MPC with one-step delay compensation

The steady-state responses at 1000 rpm are presented in this section. From top to bottom, the curves shown are torque, stator flux and switching frequency, respectively.

Figures 2.35 and 2.36 show the measured steady-state performance at 1000 rpm. It is seen that the implementation of MPC can reduce the torque ripple, but does not reduce the flux ripple. When the one-step delay is compensated, a significant decrease of torque and flux ripples can be found as well as an obvious increase of switching frequency. When the linear multiple horizon prediction is added to MPC, it can be seen that the torque and flux ripples are slightly decreased along with a limited reduction of the switching frequency.

Table 2.8 lists the torque and flux ripples of these control methods in experiment. As shown, similar conclusions can be obtained as those from Table 2.7. According to the analysis above, it can be concluded that:

(1) MPC can achieve lower torque ripple than that of DTC whilst maintaining/reducing the switching frequency as proven in both simulation and experimental tests. However, MPC's ability in flux ripple reduction is insignificant and even unstable.

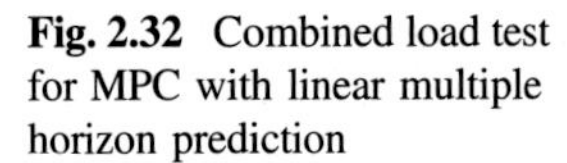
Fig. 2.32 Combined load test for MPC with linear multiple horizon prediction

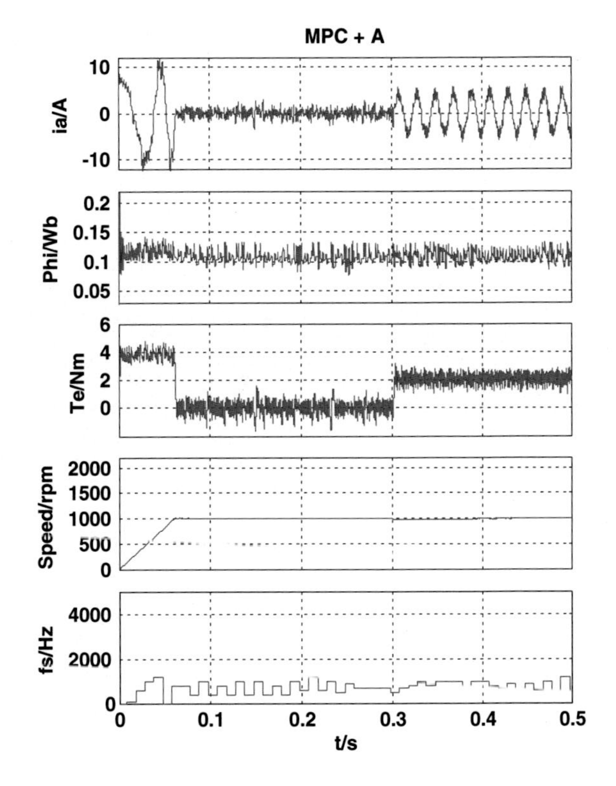

Table 2.7 Steady-state response (simulation)

Method	THD (%)	f_{av} (kHz)	ψ_{rip}(Wb)	T_{rip}(Nm)
DTC	28.83	1.5972	0.0155	0.6869
MPC	18.55	1.5692	0.0138	0.4952
MPC + comp	8.17	1.5812	0.0059	0.2253
MPC + A	15.52	0.6640	0.0090	0.5102

(2) When one-step delay is compensated, the steady-state performance of MPC in terms of torque and flux ripples reduction is significantly improved. It should be noticed that the performance improvement also comes with a remarkable switching frequency increase (two times or more).

(3) By introducing linear multiple horizon prediction to MPCs, a significant switching frequency reduction can be found as well as an obvious decrease in flux ripple. However, it comes with heavy penalty of torque ripple increasing, especially at low motor speed.

Fig. 2.33 Experimental setup of testing system: **a** overview of the testing platform and **b** front view of the PMSM and inverter control board

2.6.6 Improved MPC with Duty Ratio Optimization

There are many improvements for these control algorithms. One of them known as MPC with duty ratio optimization will be selected for the control of PM-SMC TFM. As a general algorithm, the theory and test results will be presented in this section.

Fig. 2.34 Dynamo-meter controller DSP6000

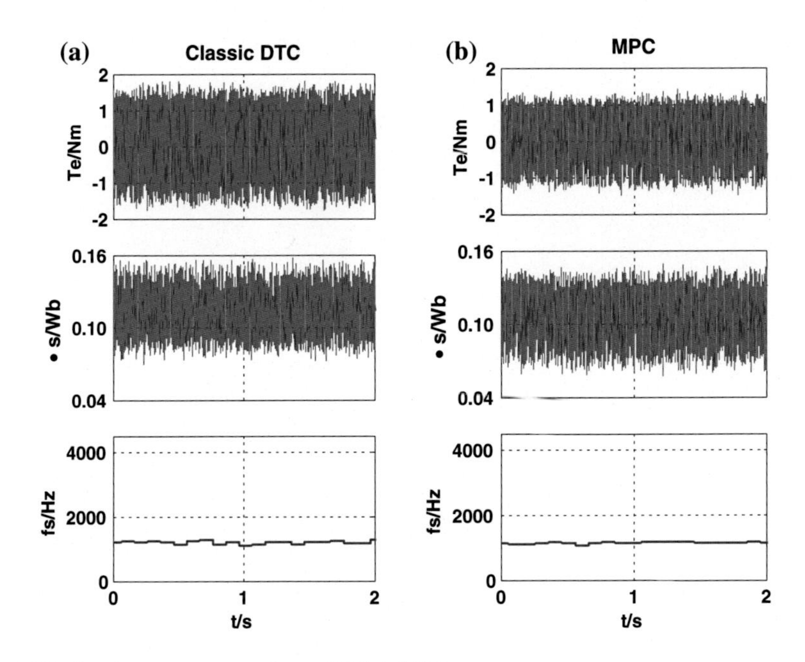

Fig. 2.35 Steady-state response for: **a** DTC, **b** MPC

In the conventional MPC, the selected voltage vector works during the whole sampling period. In many cases, it is not necessary to work for the entire period to meet the performance requirement of torque and flux. This is one of the main reasons for the torque and flux ripples. By introducing a null vector to each sampling period, the effects of voltage on torque can be adjusted to be more moderate, in order to diminish the ripples of torque and flux.

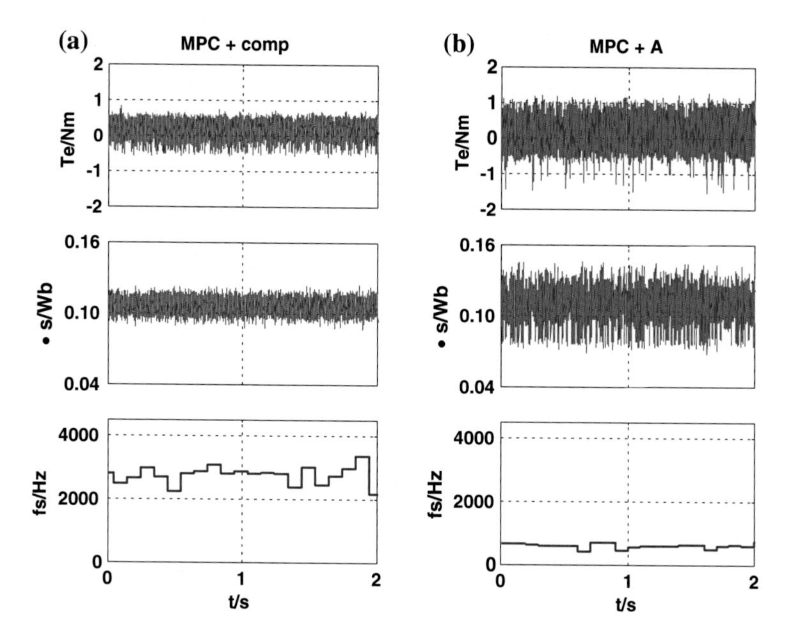

Fig. 2.36 Steady-state response for: **a** MPC with one-step delay compensation, and **b** MPC with linear multiple horizon prediction

Table 2.8 Steady-state response at 1000 rpm (experimental)

Method	f_{av} (kHz)	ψ_{rip}(Wb)	T_{rip}(Nm)
DTC	1.2129	0.0167	0.8446
MPC	1.1393	0.0173	0.6394
MPC + comp	2.7335	0.0056	0.2310
MPC + A	0.6045	0.0136	0.4460

Actually, the torque can be changed by adjusting the amplitude and time duration of u_s. The amplitude is decided by the DC bus voltage and is usually fixed, while the time duration of u_s can be varied from zero to the whole period, which is equivalent to changing the voltage vector length. The null vector only decreases the torque, while appropriate non-zero vectors can increase the torque, and it is possible to employ both null and nonzero vectors during one cycle to reduce the torque ripple. The appropriate non-zero vectors are also referred as 'active vector'. The key issue is how to determine the time duration of the two vectors, or the duty ratio of the active vector.

The expression of duty ratio for MPC is shown as follows

$$d = \left|\frac{T_e^* - T_e^{k+1}}{C_T}\right| + \left|\frac{\psi_s^* - \psi_e^{k+1}}{C_\psi}\right|, \tag{2.49}$$

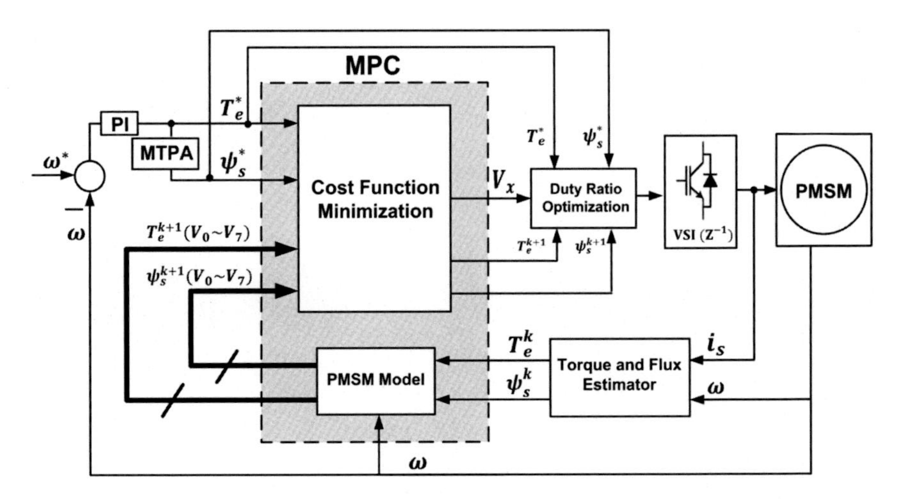

Fig. 2.37 Diagram of an improved MPC with duty ratio optimization in MATLAB/SIMULINK

where d is the duty ratio of the active voltage vector, and C_T and C_ψ are two positive parameters. The idea of this method is that the larger the difference between the reference and predicted torque values, the larger is the duty ratio value [36]. On the other hand, the lower the C_T and C_ψ values, the quicker is the dynamic response (e.g. take less time to reach the given speed), but the poorer will be the steady-state response (e.g. higher torque and flux ripples). Higher values of C_T and C_ψ could lead to better steady-state responses, but slower dynamic responses. Therefore, the determination of these values is a compromise between the steady-state and dynamic performances. Extensive simulation and experimental results have proven that the PM flux value and half-rated torque value for C_T and C_ψ can provide a good compromise between the steady state performance and the dynamic response. The block diagram of the proposed improved MPC is shown in Fig. 2.37.

2.6.7 *Numerical and Experimental Comparisons of DTC and MPC with Duty Ratio Optimization*

2.6.7.1 Numerical Simulation

The parameters of the motor and control system simulated in this section are listed in Table 2.9. Similar to the previous test example, this simulation test combines the start-up, steady-state and external load tests. The motor starts up from 0 s with several reference speeds (500 rev/min, 1000 rev/min, 1500 rev/min and 2000 rev/min). After reaching the reference speed, the motor maintains the speed for at least 0.2 s and an external load is applied at 0.3 s. Figure 2.38 shows the combined load test for one reference speed, 1000 rev/min. From top to bottom, the curves are

Table 2.9 Motor and control system parameters

Parameter	Symbol	Value
Number of pole pairs	p	3
Permanent magnet flux	ψ_f	0.1057 Wb
Stator resistance	R_s	1.8 Ω
d-axis and q-axis inductance	L_d, L_q	15 mH
DC bus voltage	V_{dc}	200 V
Inertia	J	0.002 kg · m^2
Torque constant gain	C_T	2
Flux constant gain	C_ψ	0.1
Sampling frequency	f_{sp}	5 kHz

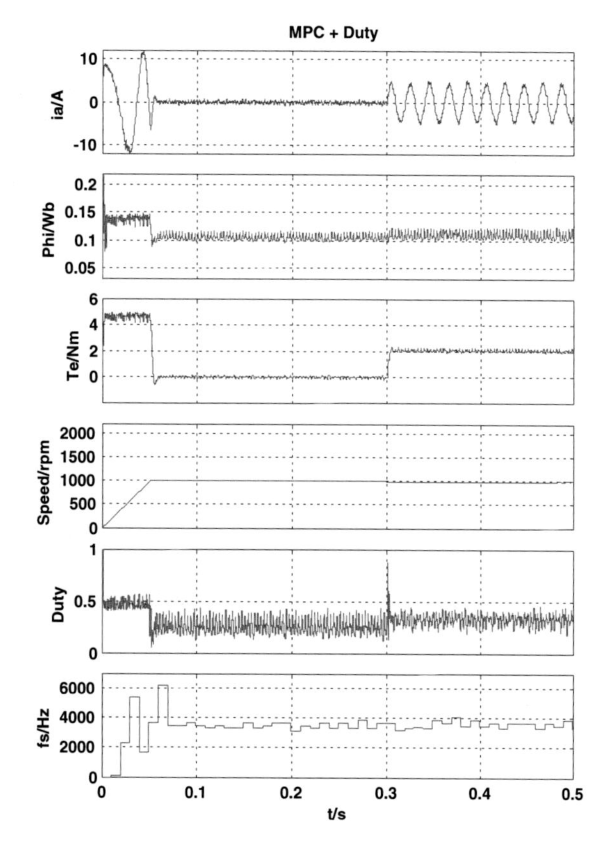

Fig. 2.38 Combined load test for MPC with duty ratio optimization at 1000 rev/min

the stator current, stator flux, torque, motor speed, and switching frequency, respectively. The test results for other speed situations can be found in [36].

It can be found that the proposed MPC scheme present very low torque and flux ripples and excellent dynamic response. The proposed MPC scheme also presents very low stator current THDs and narrow harmonic spectrums with the dominant harmonics of around 5 kHz.

2.6.7.2 Experimental Test

The experimental tests are performed on the same testing platform introduced in the last section. Figure 2.39 shows the steady-state responses at 1000 rpm for three control strategies, namely (a) DTC, (b) MPC, and (c) MPC with duty ratio optimization.

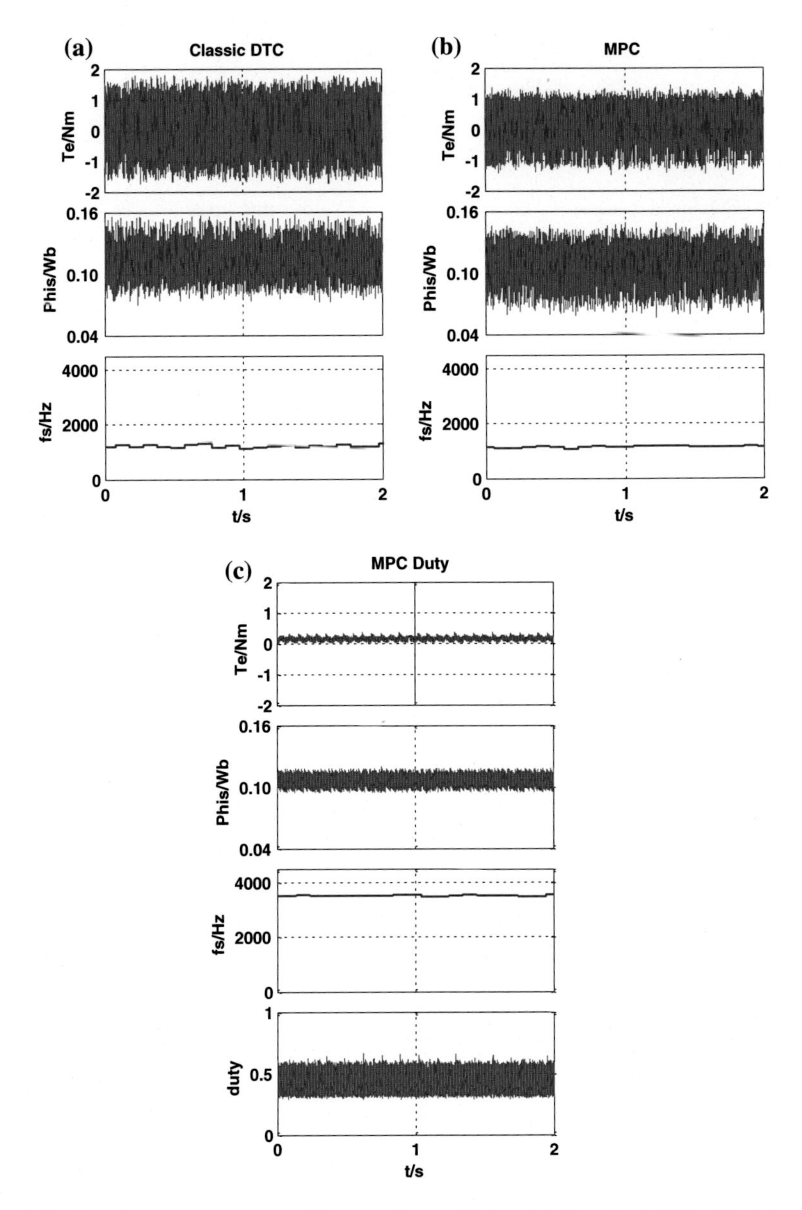

Fig. 2.39 Steady-state response at 1000 rpm for: **a** DTC, **b** MPC and **c** MPC with duty ratio optimization

It can be seen that in MPC with duty ratio optimization, the torque and flux ripples are reduced significantly compared to other methods. The duty ratio increases along with the increase in motor speed.

According to the analysis above, it can be concluded that:

(1) MPC with duty ratio optimization can achieve a better performance than DTC and original MPC in terms of torque and flux ripples reduction;
(2) Under the same system sampling frequency (5 kHz), the switching frequency of the improved method is much higher than other methods; and
(3) In DTC and MPC, the switching frequency slightly decreases along with the increase of motor speed. However, the switching frequency is almost stable in the proposed method.

More experimental results including different speed, dynamic response and data analysis can be found in [36].

2.7 Summary

This chapter presents the multi-disciplinary design analysis models and methods for electrical machines and drive systems. All the models and methods are discussed in terms of the three major parts of electrical drive systems, namely electrical machines, power electronic converters and controllers. Electromagnetic, thermal and mechanical analyses based on different models, e.g. FEM, have been investigated for the design of electrical machines with several prototypes developed in our research center. Various kinds of popular control algorithms have been described for the controller design. Several examples investigated in our previous work have been presented to show the effectiveness of the proposed models and analysis methods.

References

1. Lei G, Wang TS, Guo YG, Zhu JG, Wang SH (2014) System level design optimization methods for electrical drive systems: deterministic approach. IEEE Trans Ind Electron 61 (12):6591–6602
2. Lei G, Wang TS, Zhu JG, Guo YG, Wang SH (2015) System level design optimization method for electrical drive system: robust approach. IEEE Trans Ind Electron 62(8):4702–4713
3. Zhu JG, Guo YG, Lin ZW, Li YJ, Huang YK (2011) Development of PM transverse flux motors with soft magnetic composite cores. IEEE Trans Magn 47(10):4376–4383
4. Zhu JG, Ramsden VS (1998) Improved formulations for rotational core losses in rotating electrical machines. IEEE Trans Magn 34(4):2234–2242
5. Guo YG, Zhu JG, Lu HY, Lin ZW, Li YJ (2012) Core loss calculation for soft magnetic composite electrical machines. IEEE Trans Magn 48(11):3112–3115

6. Guo YG, Zhu JG, Lu HY, Li YJ, Jin JX (2014) Core loss computation in a permanent magnet transverse flux motor with rotating fluxes. IEEE Trans Magn 50(11). Article#: 6301004
7. Guo YG, Zhu JG, Zhong JJ, Wu W (2003) Core losses in claw pole permanent magnet machines with soft magnetic composite stators. IEEE Trans Magn 39(5):3199–3201
8. Huang YK, Zhu JG, Guo YG, Lin ZW, Hu Q (2007) Design and analysis of a high speed claw pole motor with soft magnetic composite core. IEEE Trans Magn 43(6):2492–2494
9. Huang YK, Zhu JG et al (2009) Thermal analysis of high-speed SMC motor based on thermal network and 3-D FEA with rotational core loss included. IEEE Trans Magn 45(10):4680–4683
10. Pfister P-D, Perriard Y (2010) Very-high-speed slotless permanent-magnet motors: analytical modeling, optimization, design, and torque measurement methods. IEEE Trans Ind Electron 57(1):296–303
11. Komeza K, Dems M (2012) Finite-element and analytical calculations of no-load core losses in energy-saving induction motors. IEEE Trans Ind Electron 59(7):2934–2946
12. Wang SH, Meng XJ, Guo NN, Li HB, Qiu J, Zhu JG et al (2009) Multilevel optimization for surface mounted PM machine incorporating with FEM. IEEE Trans Magn 45(10):4700–4703
13. Barcaro M, Bianchi N, Magnussen F (2012) Permanent-magnet optimization in permanent-magnet-assisted synchronous reluctance motor for a wide constant-power speed range. IEEE Trans Ind Electron 59(6):2495–2502
14. Vese I, Marignetti F, Radulescu MM (2010) Multiphysics approach to numerical modeling of a permanent-magnet tubular linear motor. IEEE Trans Ind Electron 57(1):320–326
15. Bornschlegell AS, Pelle J, Harmand S, Fasquelle A, Corriou J-P (2013) Thermal optimization of a high-power salient-pole electrical machine. IEEE Trans Ind Electron 60(5):1734–1746
16. Lee D-H, Pham TH, Ahn J-W (2013) Design and operation characteristics of four-two pole high-speed SRM for torque ripple reduction. IEEE Trans Ind Electron 60(9):3637–3643
17. Flieller D, Nguyen NK, Wira P, Sturtzer G, Abdeslam DO, Merckle J (2014) A self-learning solution for torque ripple reduction for nonsinusoidal permanent-magnet motor drives based on artificial neural networks. IEEE Trans Ind Electron 61(2):655–666
18. Hasanien HM, Abd-Rabou AS, Sakr SM (2010) Design optimization of transverse flux linear motor for weight reduction and performance improvement using response surface methodology and genetic algorithms. IEEE Trans Energy Convers 25(3):598–605
19. Hasanien HM (2011) Particle swarm design optimization of transverse flux linear motor for weight reduction and improvement of thrust force. IEEE Trans Ind Electron 58(9):4048–4056
20. Lei G, Liu CC, Guo YG, Zhu JG (2015) Multidisciplinary design analysis for PM motors with soft magnetic composite cores. IEEE Trans Magn 51(11). Article 8109704
21. Hua W, Cheng M, Zhu ZQ, Howe D (2006) Design of flux-switching permanent magnet machine considering the limitation of inverter and flux-weakening capability. In: Proceedings of 41st IAS annual meeting-industry applications conference, vol 5, pp 2403–2410
22. Liu CC, Zhu JG, Wang YH, Lei G, Guo YG, Liu XY (2014) A low-cost permanent magnet synchronous motor with SMC and ferrite PM. In: Proceedings of 17th international conference on electrical machines and systems (ICEMS), pp 397–400
23. Fei W, Luk PCK, Shen JX, Wang Y, Jin M (2012) A novel permanent-magnet flux switching machine with an outer-rotor configuration for in-wheel light traction applications. IEEE Trans Ind Appl 48(5):1496–1506
24. Guo YG (2003) Development of low cost high performance permanent magnet motors using new soft magnetic composite materials, UTS thesis (PhD)
25. Guo YG, Zhu JG, Watterson PA, Wei Wu (2006) Development of a PM transverse flux motor with soft magnetic composite core. IEEE Trans Energy Conver 21(2):426–434
26. Guo YG, Zhu JG, Watterson PA, Wei Wu (2003) Comparative study of 3-D flux electrical machines with soft magnetic composite cores. IEEE Trans Ind Appl 39(6):1696–1703
27. Guo YG, Zhu JG, Dorrell D (2009) Design and analysis of a claw pole PM motor with molded SMC core. IEEE Trans Magn 45(10):582–4585
28. Lei G, Shao KR, Guo YG, Zhu JG (2012) Multi-objective sequential optimization method for the design of industrial electromagnetic devices. IEEE Trans Magn 48(11):4538–4541

29. Lei G, Guo YG, Zhu JG et al (2012) System level six sigma robust optimization of a drive system with PM transverse flux machine. IEEE Trans Magn 48(2):923–926
30. Lei G, Zhu JG, Guo YG, Hu JF, Xu W, Shao KR (2013) Robust design optimization of PM-SMC motors for Six Sigma quality manufacturing. IEEE Trans Magn 49(7):3953–3956
31. Lei G, Zhu JG, Guo YG, Shao KR, Xu W (2014) Multiobjective sequential design optimization of PM-SMC motors for six sigma quality manufacturing. IEEE Trans Magn, 50 (2). Article 7017704
32. Liu CC, Zhu JG, Wang YH, Guo YG, Lei G, Liu XY (2015) Development of a low-cost double rotor axial flux motor with soft magnetic composite and ferrite permanent magnet materials. J Appl Phys, 117(17). Article # 17B507
33. Teng QF, Zhu JG, Wang TS, Lei G (2012) Fault tolerant direct torque control of three-phase permanent magnet synchronous motors. WSEAS Trans Syst 8(11):465–476
34. Teng QF, Bai J, Zhu JG, Sun Y (2013) Fault tolerant model predictive control of three-phase permanent magnet synchronous motors. WSEAS Trans Syst 12(8):385–397
35. Wang Y (2011) Investigation of rotor position detection schemes for PMSM drives based on analytical machine model incorporating nonlinear saliencies, UTS thesis (PhD)
36. Wang TS (2013) Model predictive torque control of PMSM with duty ratio optimization for torque ripple reduction, UTS thesis (Master degree)
37. Kim SY, Lee W, Rho MS, Park SY (2010) Effective dead-time compensation using a simple vectorial disturbance estimator in PMSM drives. IEEE Trans Ind Electron 57(5):1609–1614
38. Lee J, Hong J, Nam K, Ortega R, Praly L, Astolfi A (2010) Sensorless control of surface-mount permanent-magnet synchronous motors based on a nonlinear observer. IEEE Trans Power Electron 25(2):290–297
39. Genduso F, Miceli R, Rando C, Galluzzo GR (2010) Back EMF sensorless-control algorithm for high-dynamic performance PMSM. IEEE Trans Ind Electron 57(6):2092–2100
40. Takahashi I, Noguchi T (1986) A new quick-response and high-efficiency control strategy of an induction motor. IEEE Trans Ind Appl 22(5):820–827
41. Depenbrock M (1988) Direct self-control (DSC) of inverter-fed induction machine. IEEE Trans Power Electron 3(4):420–429
42. Buja GS, Kazmierkowski MP (2004) Direct torque control of PWM inverter-fed AC motors-A survey. IEEE Trans Ind Electron 51(4):744–757
43. Lai YS, Chen JH (2001) A new approach to direct torque control of induction motor drives for constant inverter switching frequency and torque ripple reduction. IEEE Trans Energy Convers 16(3):220–227
44. Lascu C, Trzynadlowski A (2004) A sensorless hybrid DTC drive for high-volume low-cost applications. IEEE Trans Ind Electron 51(5):1048–1055
45. Zhang Y, Zhu J, Xu W, Hu J, Dorrell DG, Zhao Z (2010) Speed sensorless stator flux oriented control of three-level inverter-fed induction motor drive based on fuzzy logic and sliding mode control. In: Proceedings of 36th IEEE IECON, pp 2926–293
46. Zhang Y, Zhu J (2011) Direct torque control of permanent magnet synchronous motor with reduced torque ripple and commutation frequency. IEEE Trans Power Electron 26(1):235–248
47. Zhang Y, Zhu J (2011) A novel duty cycle control strategy to reduce both torque and flux ripples for DTC of permanent magnet synchronous motor drives with switching frequency reduction. IEEE Trans Power Electron 26(10):3055–3067
48. Zhang Y, Zhu J, Xu W, Guo Y (2011) A simple method to reduce torque ripple in direct torque-controlled permanent-magnet synchronous motor by using vectors with variable amplitude and angle. IEEE Trans Ind Electron 58(7):2848–2859
49. Wang TS, Zhu JG, Zhang YC (2011) Model predictive torque control for PMSM with duty ratio optimization. In Proceedings of 2011 international conference on electrical machines and systems (ICEMS), pp 1–5, 20–23 August 2011
50. Miranda H, Cortes P, Yuz J, Rodriguez J (2009) Predictive torque control of induction machines based on state-space models. IEEE Trans Ind Electron 56(6):1916–1924
51. Geyer T, Papafotiou G, Morari M (2009) Model predictive direct torque control—Part I: Concept, algorithm, and analysis. IEEE Trans Ind Electron 56(6):1894–1905

52. Kouro S, Cortes P, Vargas R, Ammann U, Rodriguez J (2009) Model predictive control—a simple and powerful method to control power converters. IEEE Trans Ind Electron 56 (6):1826–1838
53. Morel F, Retif J-M, Lin-Shi X, Valentin C (2008) Permanent magnet synchronous machine hybrid torque control. IEEE Trans Ind Electron 55(2):501–511
54. Drobnic K, Nemec M, Nedeljkovic D, Ambrozic V (2009) Predictive direct control applied to AC drives and active power filter. IEEE Trans Ind Electron 56(6):1884–1893
55. Zhang Y, Xie W (2014) Low complexity model predictive control-single vector-based approach. IEEE Trans Power Electron 29(10):5532–5541
56. Zhang Y, Qu C (2015) Model predictive direct power control of PWM rectifiers under unbalanced network conditions. IEEE Trans Ind Electron 62(7):4011–4022

Chapter 3
Optimization Methods

Abstract Optimization is an art of searching the best one/ones among a great number of feasible solutions. The main optimization target of electromagnetic devices and systems including electrical machines and drive systems is to determine a set of parameters involving material, topology and structural parameters to satisfy certain design specifications and constraints, such as output power, efficiency, volume, and cost. Engineers have been using optimization methods to optimize the designs of electromagnetic devices, components and systems for decades. This chapter aims to presents the optimization methods commonly used in the field of electrical machines and drive systems, as well as computational electromagnetics. Classic and modern intelligent optimization algorithms will be discussed firstly, followed by the multi-objective optimization algorithms. Four kinds of approximate models will be described, and the modelling methods will be discussed with two numerical examples.

Keywords Optimization methods · Intelligent optimization algorithms · Approximate models · Multi-objective optimization algorithms

3.1 Introduction

In Chap. 2, the design fundamentals, and various design analysis models for electrical machines and drive systems have been investigated in terms of different disciplines or subject domains, such as the analytical models or methods for electromagnetic and thermal analyses, magnetic circuit model for electromagnetic analysis, finite element model (FEM) for all electromagnetic, thermal and mechanical analyses, and field oriented control (FOC), direct torque control (DTC), and model predictive control (MPC) algorithms for the control systems. While some of them are physical analysis models which can reveal the basic operational principles of the electrical machines and drive systems, the FEM is a kind of numerical analysis model, which is widely used in the design optimization of electrical machines to get a further understanding and illustrations for the field

G. Lei et al., *Multidisciplinary Design Optimization Methods for Electrical Machines and Drive Systems*, Power Systems,
DOI 10.1007/978-3-662-49271-0_3

analysis. All these design analysis models can be employed for the performance evaluation of electrical machines and drive systems.

On the other hand, solving an optimization problem consists of two main issues: definition of optimization models and selection/development of (new) optimization methods. Generally, an optimization model for a single-objective with m constraints can be defined as

$$\begin{array}{ll} \min: & f(\mathbf{x}) \\ \text{s.t.} & g_i(\mathbf{x}) \le 0,\ i = 1, \ldots, m, \\ & \mathbf{x}_l \le \mathbf{x} \le \mathbf{x}_u \end{array} \tag{3.1}$$

where $\mathbf{x}$, f and g are the design parameter vector, objectives and constraints, respectively, $\mathbf{x}_l$ and $\mathbf{x}_u$ the lower and upper boundaries of $\mathbf{x}$, respectively, and m is the number of constraints. Typical optimization objectives for electrical machines and drive systems are the minimization of cost, cogging torque, torque ripples, overshoot, and maximization of output power and efficiency. Popular constraints are volume, mass, current density and temperature rises [1].

Theoretically, the optimization model (3.1) is always a strongly-constrained, highly- nonlinear and high-dimensional problem for electrical machines and drive systems. Many kinds of optimization algorithms have been employed to find the optimum for the above equation, such as the sequential quadratic programming algorithm, genetic algorithm (GA), differential evolution algorithm (DEA), and particle swarm optimization (PSO) algorithm [2–5]. Section 3.2 presents an overview for the classical and modern optimization algorithms for solving (3.1).

On the other hand, from the perspective of practical engineering applications, the design optimization of electrical machines is actually a multi-objective problem as there are many objectives which can be defined and different objectives can be selected for different applications. For example, for applications in home appliances, such as washing machines and refrigerators, the motor price and output power may be the two most important issues; while for applications in hybrid electric vehicles, the volume, power density and torque ripple are very important. Therefore, multi-objective optimization design problems of electrical machines as well as other electromagnetic devices have become a topic of much interest recently [6–8].

Generally, a multi-objective optimization model with p objectives and m constraints can be defined as

$$\begin{array}{ll} \min: & \{f_1(\mathbf{x}), f_2(\mathbf{x}), \ldots f_p(\mathbf{x})\} \\ \text{s.t.} & g_i(\mathbf{x}) \le 0,\ i = 1, \ldots, m, \\ & \mathbf{x}_l \le \mathbf{x} \le \mathbf{x}_u \end{array} \tag{3.2}$$

The solutions of (3.2) are often illustrated by Pareto optimal figure, which can be obtained by using multi-objective optimization algorithms. Many multi-objective optimization algorithms developed in the field of evolutionary computation have been introduced to the design of electrical machines, such as multi-objective genetic algorithm (MOGA), non-dominated sorting genetic algorithm (NSGA) and NSGA II, multi-objective particle swarm optimization (MPSO) algorithm [6–8]. Section 3.3 briefly describes several popular multi-objective optimization algorithms employed in the design of electrical machines.

The above optimization methods (physical models or FEMs plus optimization algorithms) can be regarded as direct optimization method. Though they are always of high accuracy, the optimization efficiency is not good for many situations due to the high nonlinearity of the problem, particularly for the FEMs. Approximate models (or surrogate models) present an alternative way for the optimization to increase the optimization efficiency. Section 3.4 presents a summary for four kinds of widely used approximate models, response surface model (RSM), radial basis function (RBF) model, Kriging model and artificial neural network (ANN) model. Section 3.5 presents an introduction of construction and verification of approximation with example analyses of two classical test functions, followed by the summary Sect. 3.6.

3.2 Optimization Algorithms

3.2.1 Classic Optimization Algorithms

Many kinds of classic optimization algorithms have been introduced to solve the constrained and nonlinear optimization problem (3.1). Some are gradient-based algorithms, such as conjugate gradient algorithm, sequential quadratic programming algorithm and augmented Lagrange multiplier method [5, 9–12]. Generally, the first or second order derivative or Hessians matrix is required in the implementation. To use these algorithms efficiently, there are several constraints, such as

(a) The objective functions should be continuous and derivable;
(b) The objective functions and constraints can be expressed analytically; and
(c) The constrained optimization models have to be converted to unconstrained forms for some initial gradient-based algorithms, e.g. the conjugate gradient algorithm.

Analytical models or methods for electromagnetic, thermal and other disciplinary analyses should be constructed before the optimization. However, many analysis models for electrical machines are based on FEM, and there is no analytical expression for the optimization model. Therefore, various intelligent optimization algorithms using non-analytical machine models have been employed, such as those based on the GA and PSO algorithms.

3.2.2 Modern Intelligent Algorithms

In the past several decades, a number of innovative intelligent algorithms, such as the evolutionary algorithms including GA and DEA, PSO algorithms, immune algorithm, and ant colony algorithm, have been developed and employed for design optimization problems [13–18]. Several evolutionary algorithms will be presented in this chapter due to the wide usage of them. PSO will also be introduced in this section as an example of different kind of optimization algorithms.

The evolutionary algorithms (EAs) are a kind of heuristic optimization algorithms, which use techniques inspired by mechanisms from biological evolution such as reproduction, mutation, recombination, natural selection and survival of the fittest to find an optimal configuration for a specific system. There are four main branches during the development of EAs: (a) GAs, (b) evolution programming, (c) evolution strategy, and (d) differential evolution. The general flowchart for EAs is illustrated in Fig. 3.1.

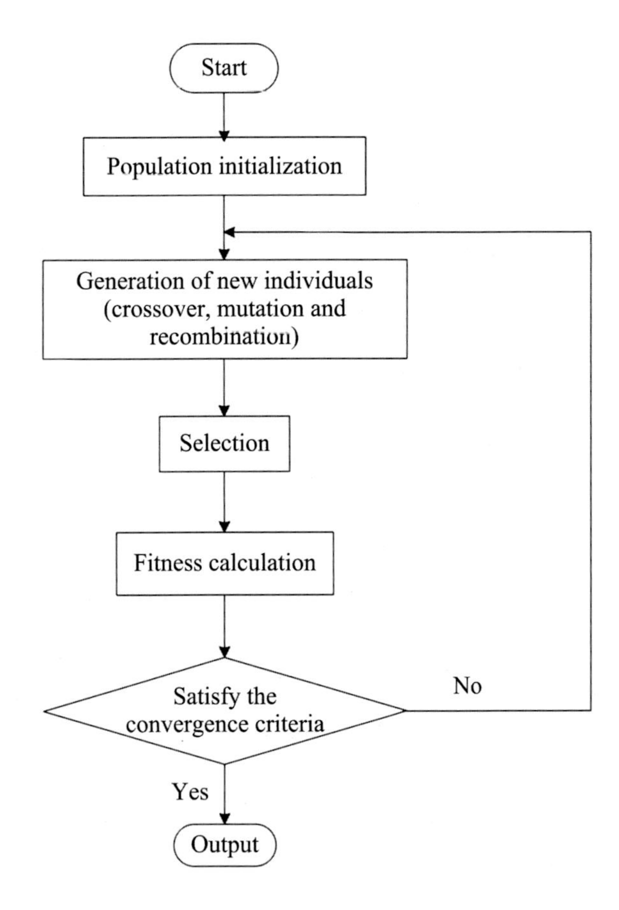

Fig. 3.1 Flowchart for general EAs

In the flowchart of EAs, fitness calculation is related to the objective functions, which is used to evaluate the performance of each individual in the (initial) population. Then, the algorithm parameters should be defined prior to the implementation of the optimization, such as the population size, crossover, mutation, selection and recombination factors, maximal iteration number, and convergence criteria. Through a broad research, it is found that EAs have the following merits:

(a) They are global optimization methods;
(b) They can be applied to almost any optimization problems and scale well to higher dimensional problems;
(c) They are robust in terms of noisy evaluation functions;
(d) They are conceptually simple and can easily be adjusted to the problem at hand; Almost any aspect of the algorithm may be changed and customized; and
(e) EAs have strong parallel searching capability as evolution is a highly parallel process.

Several EAs will be introduced as follows with more details, and several of them will be used as optimization algorithms in this book.

3.2.2.1 GAs

GAs have been employed in science and engineering as adaptive intelligent algorithms for solving practical problems. They are inspired by Darwin's theory about evolution. Solution to a problem provided by GAs is evolved. Figure 3.2 illustrates a general optimization flowchart of GAs. As shown, the algorithm is started with a set of population (represented by chromosomes). Solutions from one population (known as parent) are taken and used to form a new population (known as offspring or children) by three genetic operations, crossover, mutation and selection. Solutions which are used to generate new solutions (offspring or children) are selected in terms of their fitness, which means that the more suitable they are, the more opportunities they have to reproduce in the evolution process [14, 19, 20]. This is repeated until some conditions or criteria are satisfied, for example the maximal iteration number. The outline of the basic GA is listed as follows.

(a) Start—Generate initial population of NP chromosomes;
(b) Fitness—Evaluate the fitness $f(x)$ of each chromosome x in the initial population;
(c) New population—Create a new population by repeating following steps;

- Selection—Select two parent chromosomes from a population in terms of their fitness (the better fitness, the bigger chance to be selected);
- Crossover—Form a new offspring with a crossover probability over the parents;
- Mutation—Mutate new offspring at each locus (position in chromosome) with a mutation probability;

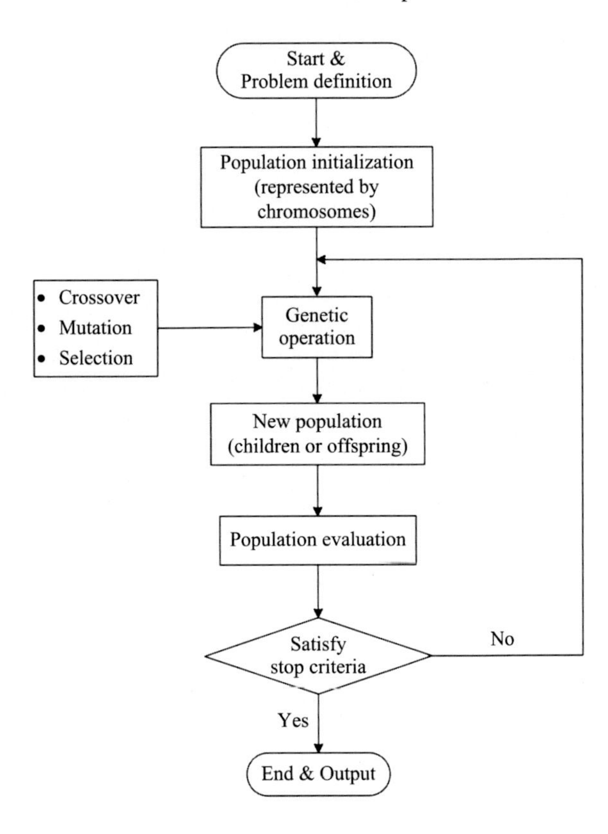

Fig. 3.2 Flowchart for general GAs

(d) Replace—Use new generated population for a further run of algorithm; and
(e) Test—If the end condition is satisfied, stop, and return the best solution in current population. Otherwise, go to step (b) and do the optimization loop till convergence.

In the implementation, there are several algorithm parameters needed to define. They are the population size (*NP*), crossover probability (P_c), mutation probability (P_m) and determination of the type of selection strategies. Generally, the population size can be defined as 5–10 times of the problem dimension, P_c = 0.6–1.0, and P_m = 0.005–0.05. For the selection operation (regarding the problem of how to select parents for crossover), this can be done in many ways. A popular one is to select the better parents (assuming that the better parents will produce better offspring), and this is generally called elitism select strategy. It means that at least one best solution is copied without changes to a new population, so that the best solution found can survive to the end of run. Other selection strategies are roulette wheel selection and rank selection methods [19, 20].

GAs have many advantages. For example, GAs work on the chromosome, which is an encoded version of potential solutions' parameters, rather than the parameters themselves. On the other hand, they use fitness score, which is obtained from objective functions, without other derivative or auxiliary information. Thus, they have the ability to avoid to be trapped in local optimal solution unlike the traditional methods, which search from a single point.

3.2.2.2 DEA

DEA is a relatively new evolutionary optimization algorithm [21–23]. Many studies demonstrated that DEA converges fast and is robust, simple to implement, requiring only a few control parameters. The procedure of DEA is almost the same as that of the GA whose main process has mutation, crossover, and selection. The main difference between DEA and GA lies in the mutation process.

Figure 3.3 shows the optimization flowchart of DEA. The implementation of DEA consists of the following five main steps:

Step 1: Population initialization

Assume that $\{\mathbf{x}_i^t, i = 1, 2, \ldots, NP\}$ is the population, where NP is the population size. The initial population can be defined as

$$x_{ji}^0 = x_j^{(L)} + rand_{ji}[0, 1] \cdot (x_j^{(U)} - x_j^{(L)}), j = 1, 2, \ldots, D \tag{3.3}$$

where $x_j^{(L)}$ and $x_j^{(U)}$ are the lower and upper boundaries of x. In details, it can be expressed as

$$\begin{cases} x_{1i}^0 = x_1^{(L)} + rand_{1i}[0, 1] \cdot (x_1^{(U)} - x_1^{(L)}) \\ x_{2i}^0 = x_2^{(L)} + rand_{2i}[0, 1] \cdot (x_2^{(U)} - x_2^{(L)}) \\ \vdots \\ x_{Di}^0 = x_D^{(L)} + rand_{Di}[0, 1] \cdot (x_D^{(U)} - x_D^{(L)}) \end{cases} \tag{3.4}$$

Step 2: Mutation process

Assume that

$$\mathbf{v}_i^{t+i} = \mathbf{x}_{r1}^t + F(\mathbf{x}_{r2}^t - \mathbf{x}_{r3}^t), \tag{3.5}$$

where $r1$, $r2$, and $r3$ are three different numbers in [1, NP], that are different from i, and $F \in [0, 2]$ is the mutation factor. This mutation method is called as DE/rand/1. There are several other situations:

$$\mathrm{DE/best/1}: \mathbf{v}_i^{t+i} = \mathbf{x}_{best}^t + F(\mathbf{x}_{r2}^t - \mathbf{x}_{r3}^t) \tag{3.6}$$

$$\mathrm{DE/best/2}: \mathbf{v}_i^{t+i} = \mathbf{x}_{best}^t + F(\mathbf{x}_{r2}^t - \mathbf{x}_{r3}^t + \mathbf{x}_{r4}^t - \mathbf{x}_{r5}^t) \tag{3.7}$$

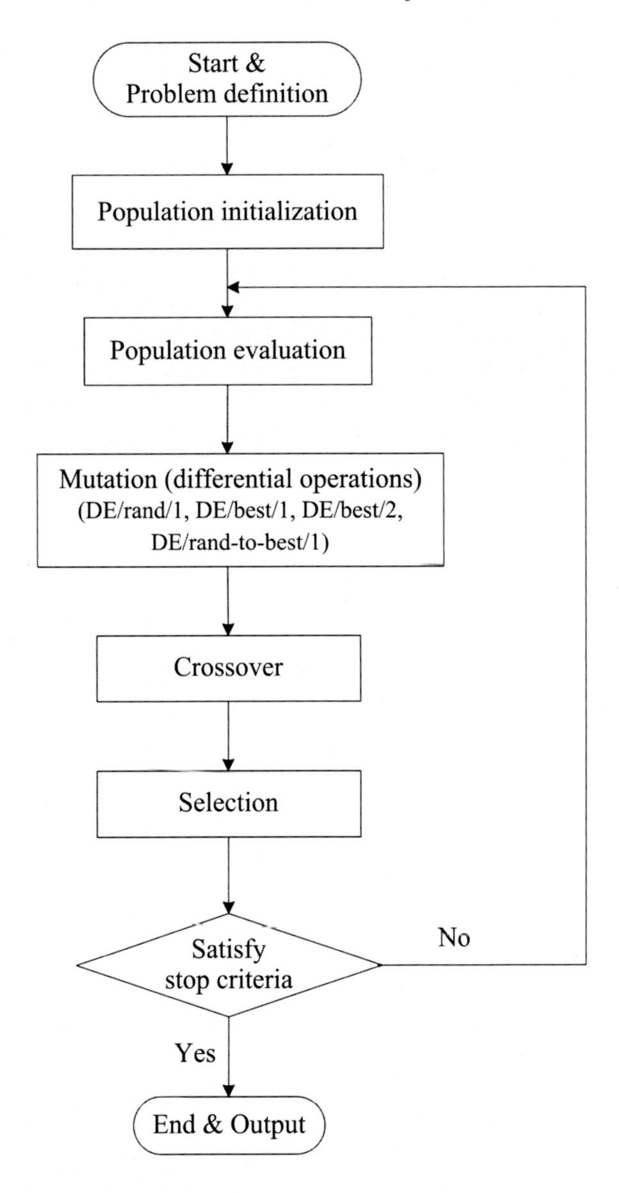

Fig. 3.3 Flowchart for general DEA

$$\text{DE/rand} - \text{to} - \text{best/1} : \mathbf{v}_i^{t+i} = \mathbf{x}_{best}^t + F_1(\mathbf{x}_{r2}^t - \mathbf{x}_{r3}^t) + F_2(\mathbf{x}_{best}^t - \mathbf{x}_{r1}^t) \tag{3.8}$$

where $r4$ and $r5$ are two different numbers in [1, NP], and subscript "best" means the best one in the iteration t.

Step 3: Crossover process

$$u_{ji}^{t+1} = \begin{cases} v_{ji}^{t+1}, & rand[0,1] \leq CR \\ x_{ji}^{t}, & \text{others} \end{cases} \tag{3.9}$$

where $CR \in [0, 1]$ is the crossover factor, and $j \in \{1, 2, \ldots, D\}$.

Step 4: Selection process

$$\mathbf{x}_i^{t+1} = \begin{cases} \mathbf{u}_i^{t+1}, & f(\mathbf{u}_i^{t+1}) < f(\mathbf{x}_i^t) \\ \mathbf{x}_i^t, & \text{others} \end{cases} \tag{3.10}$$

Step 5: Test process

If the population $\mathbf{x}_i^{t+1}$ meets the criteria, stop the optimization, and output the best one in $\mathbf{x}_i^{t+1}$ as the optimal solution. Otherwise, go to step 2 and do the optimization loop again till convergence.

From the above discussion, it can be seen that there are only 3 algorithm parameters in the DEA. They are population size (*NP*), mutation factor (*F*), and crossover factor (*CR*).

3.2.2.3 EDA

Estimation of distribution algorithms (EDAs) are a class of evolution algorithms based on probability model, sometimes known as probabilistic model-based GAs, which are an outgrowth of GAs. Figure 3.4 shows the comparison of the main flowcharts of GAs and EDAs. As shown, the main difference is the generation methods of the new population. The new population and final solutions of EDAs are obtained by learning and sampling statistically the probability distribution of the best individuals of the population in each iteration of the algorithm. The genetic operators, such as crossover and mutation, used in GAs are not required for this process in EDAs [24–27]. Therefore, EDAs have introduced a new paradigm for evolutionary computation without using the conventional evolutionary operators, and have become a hot topic in the field of evolutionary computation recently.

The most important issue in EDAs is the construction method of the probability model as shown in Fig. 3.4. According to the complexity of probability models for learning the interdependencies between the variables, a number of EDAs have been developed in terms of the interactions between parameters, namely dependency-free, bivariate dependencies, and multivariate dependencies. The popular one is the Bayesian optimization algorithm, which uses a probability graph model based on Bayesian network model to handle the interactions between different parameters. More information can be found in references [24–27].

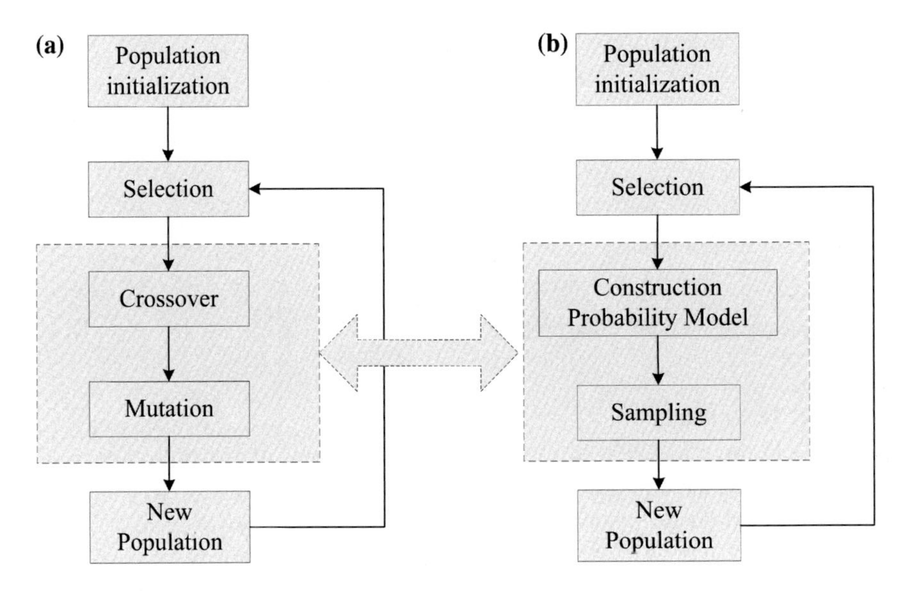

Fig. 3.4 Comparison of GAs and EDAs, **a** GA, **b** EDA

3.2.2.4 PSO

The PSO algorithm is an evolutionary algorithm that simulates the movement of flocks of birds. In this algorithm, a population of individuals (known as particles) updates their movements to reach the target point (the optimum) by continuously receiving information from other members of the flocks [11]. In the classical PSO, the *n*th particle velocity and position are updated by

$$\mathbf{v}_i^{t+1} = w\mathbf{v}_i^t + c_1 r_1(\mathbf{p}_i - \mathbf{x}_i^t) + c_2 r_2(\mathbf{p}_g - \mathbf{x}_i^t) \tag{3.11}$$

$$\mathbf{x}_{Di}^{t+1} = \mathbf{x}_{Di}^t + \alpha \mathbf{v}_{Di}^{t+1} \tag{3.12}$$

where w is the inertial weight factor, subscript D the dimension of parameter, $\mathbf{p}_i$ the local best vector of the tth particle, $\mathbf{p}_g$ the global best vector, c_1 and c_2 are adjustable social factors, r_1 and r_2 random numbers between 0 and 1, respectively, and α is the time step.

Figure 3.5 shows a flowchart of the PSO algorithm. The PSO algorithm has been used in many applications and has had many improvements. Compared with GAs, PSO is very much similar in many aspects. It is also a kind of evolutionary technique with its algorithm starting with a group of a randomly generated population, using fitness value to evaluate the population, updating the population, and searching for the optimum with random techniques.

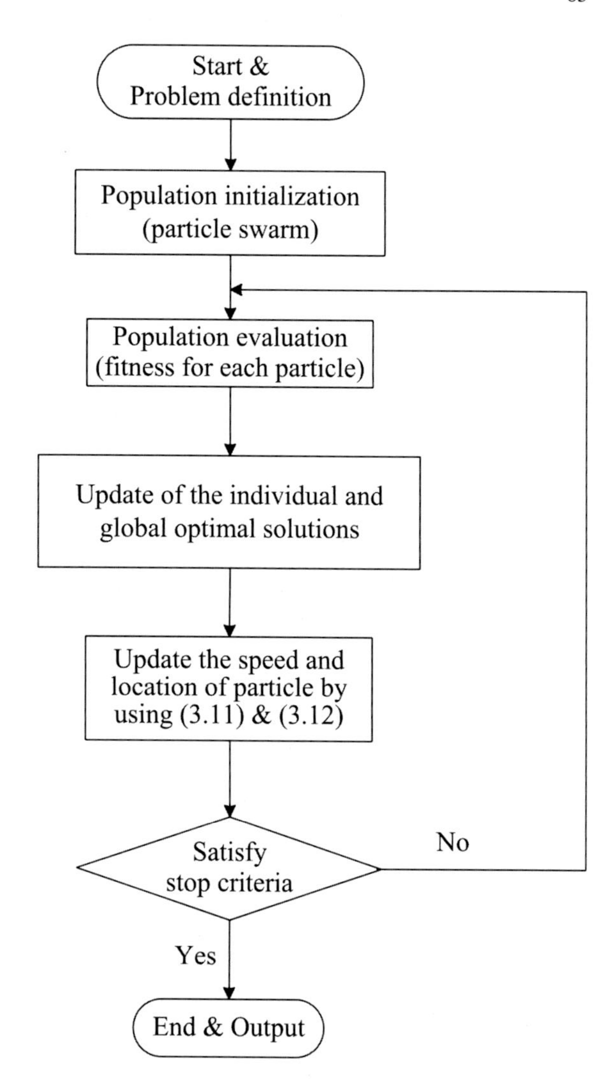

Fig. 3.5 Flowchart of PSO algorithm

However, PSO does not have genetic operators like crossover and mutation. Particles update themselves with the internal velocity. They also have memory, which is important to the algorithm.

The information sharing mechanism in PSO is also significantly different from that of GAs. In GAs, the chromosomes share information with each other, so that the whole population moves like a group towards an optimal area. In PSO, only $\mathbf{p}_g$ and $\mathbf{p}_i$ give out the information to others, which is a one-way information sharing mechanism. The evolution only looks for the best solution. Unlike GAs, in PSO, all the particles tend to converge to the best solution quickly even in the local version in most cases [11, 28–30].

3.3 Multi-objective Optimization Algorithms

3.3.1 Introduction to Pareto Optimal Solution

There are many kinds of multi-objective design optimization problems in the design of electrical machines and other electromagnetic devices [7, 8, 30–35]. Theoretically, the objectives in multi-objective optimization problems are always conflicting. The improvement of an objective may result in performance decrease of the other objectives. For example, the material cost and output power are two important issues for designing the transverse flux machines. The improvement of output power is often accompanied by the increase of material cost [7]. Therefore, it is always impossible to achieve the optimum for each of these objectives, and the corresponding optimal solutions are actually a compromise between these objectives by making the objectives close to their optimums as much as possible. The corresponding optimal solutions are called the Pareto optimal solutions. Theoretically, the Pareto solutions are only acceptable solutions or non-inferior solutions. The number of these solutions may be very large or even infinite.

There are several conceptions which are widely mentioned in the multi-objective optimization to define the Pareto optimal solutions [8].

Definition 1 Given two vectors, $\mathbf{x}, \mathbf{y} \in R^q$, we say that $\mathbf{x} \leq \mathbf{y}$ if $x_i \leq y_i$ for $i = 1,\ldots,$ q, and that $\mathbf{x}$ dominates $\mathbf{y}$ (denoted by $\mathbf{x} \prec \mathbf{y}$) if $\mathbf{x} \leq \mathbf{y}$ and $\mathbf{x} \neq \mathbf{y}$.

Figure 3.6 shows a particular case of the dominance relation in the presence of two objective functions for a minimization situation.

Definition 2 We say that a vector of decision variables $\mathbf{x} \in \mathrm{X}$ is non-dominated in X, if there does not exist another $\mathbf{x}' \in \mathrm{X}$ such that $f(\mathbf{x}') \prec f(\mathbf{x})$.

Definition 3 We say that a vector of decision variables $\mathbf{x}^* \in \mathbb{F}$ ($\mathbb{F}$ is the feasible region) is Pareto-optimal if it is non-dominated in terms of $\mathbb{F}$.

Definition 4 The Pareto optimal set is defined by

$$P^* = \{\mathbf{x} \in \mathbb{F} | \mathbf{x} \text{ is Pareto-optimal}\}$$

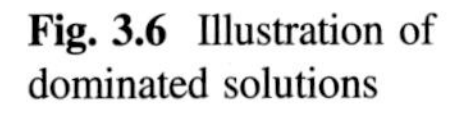
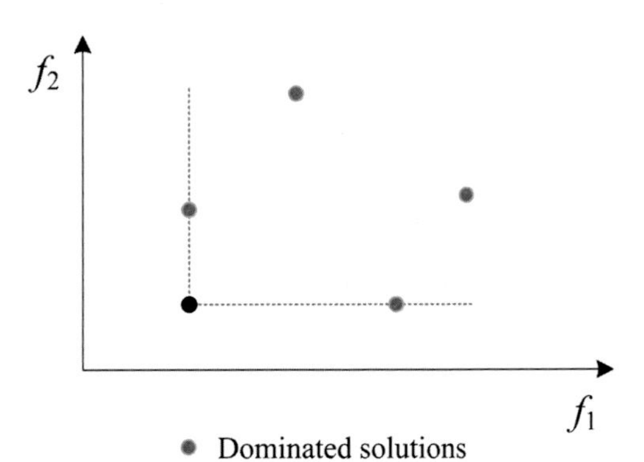

Fig. 3.6 Illustration of dominated solutions

Definition 5 The Pareto front is defined by:

$$P\mathbb{F}^* = \{f(\mathbf{x}) \in R^q | \mathbf{x} \in P^*\}$$

Figure 3.7 shows a particular case of the Pareto front in the presence of two objective functions [8].

Different from the single objective optimization algorithms, the multi-objective optimization algorithms have to provide a set of non-inferior solutions with large population, and this set approaches the front of the global Pareto optimal solutions. Those solutions should be uniformly distributed at the front of Pareto solutions as much as possible.

Based on these basic principles, a number of multi-objective optimization algorithms have been developed in the field of evolutionary computation and have been employed for the design optimization of multi-objective problems, such as MOGA, NSGA and its improvement NSGA II, MPSO algorithm [6, 36–40]. Three of them, MOGA, NSGA II and MPSO will be introduced in the following sections.

3.3.2 MOGA

Figure 3.8 illustrates a general flowchart for MOGA based on multi-objective ranking method. The multi-objective ranking method is one of the methods for evaluating the multi-objectives. Figure 3.9 shows an example of ranking in two objectives. The rank of an individual is determined as $1 + NP$ when it is dominated by other NP individuals [36, 37].

Since there is a toolbox for GA and MOGA in Matlab, it is ready to implement for practical problems. The following is an example.

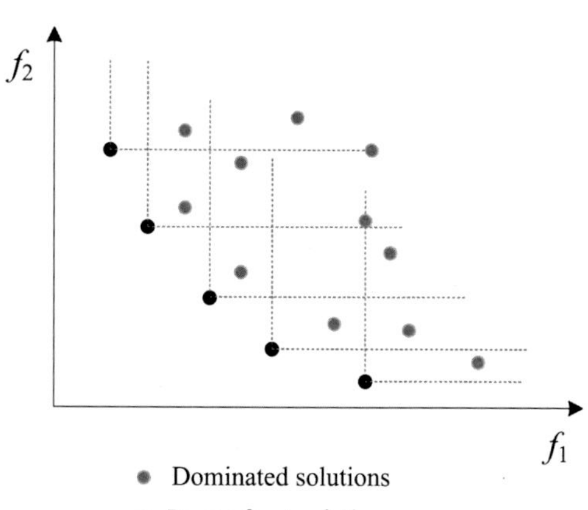

Fig. 3.7 Illustration of Pareto optimal solutions

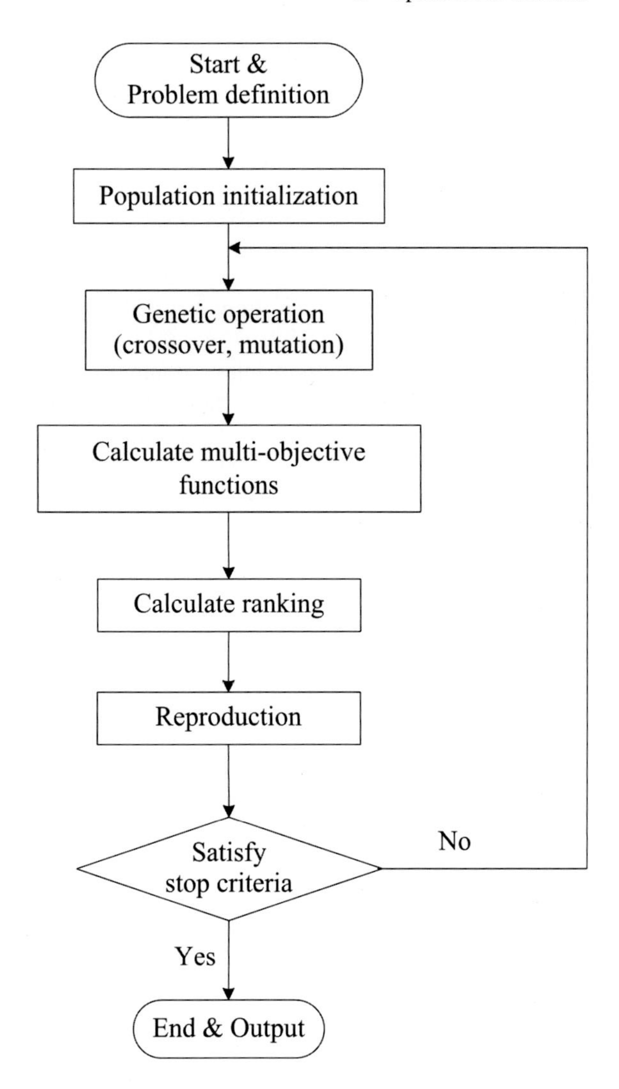

Fig. 3.8 Flowchart of MOGA

The Poloni (POL) function

$$\min: \begin{cases} f_1(x_1, x_2) = 1 + (A_1 - B_1)^2 + (A_2 - B_2)^2 \\ f_2(x_1, x_2) = (x_1 + 3)^2 + (x_2 + 1)^2 \end{cases}, \tag{3.13}$$

$$\begin{aligned} A_1 &= 0.5 \sin 1 - 2 \cos 1 + \sin 2 - 1.5 \cos 2 \\ A_2 &= 1.5 \sin 1 - \cos 1 + 2 \sin 2 - 0.5 \cos 2 \\ B_1 &= 0.5 \sin x_1 - 2 \cos x_1 + \sin x_2 - 1.5 \cos x_2 \\ B_2 &= 1.5 \sin x_1 - \cos x_1 + 2 \sin x_2 - 0.5 \cos x_2 \\ &-\pi \leq x_1, x_2 \leq \pi \end{aligned}.$$

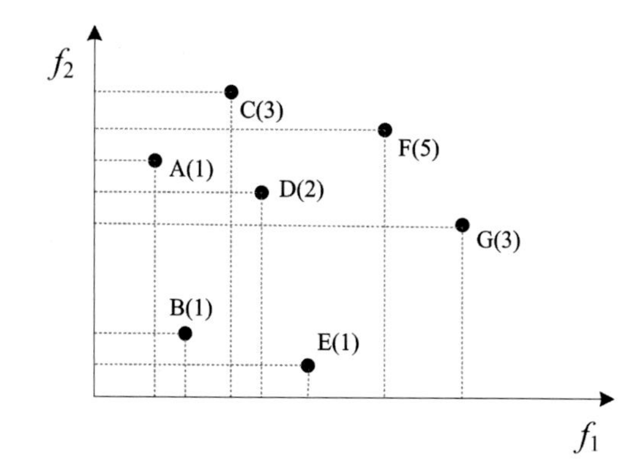

Fig. 3.9 Illustration of Multiobjective ranking method

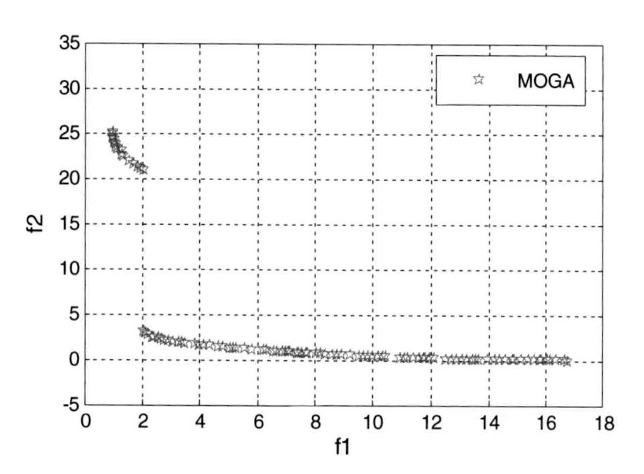

Fig. 3.10 Pareto optimal solutions for POL function by using MOGA

is a classic test function for multi-objective optimization methods as its Pareto optimal solutions are not continuous and non-convex [7, 41].

Figure 3.10 illustrates the Pareto optimal solutions obtained from MOGA. As shown, the Pareto solutions of this function are divided into two parts. It is not continuous on the whole region. This function will be used to verify the efficiency of the new proposed multi-objective optimization method in the next chapter.

3.3.3 NSGA and NSGA II

NSGA stands for non-dominated sorting genetic algorithm, which was first presented by Srinivas and Deb in 1994 [42, 43]. In this NSGA, a new method was presented to classify individuals in layers before the selection is performed. Individuals of the first layer have the highest fitness while the members of the last layer have the smallest fitness. Individuals from the first layer produce more copies than other layers in the next generation.

The NSGA II is an improved version of NSGA. It is one of the most efficient and famous multi-objective evolutionary algorithms and has been widely applied in many kinds of engineering multi-objective optimization problems. Figure 3.11 shows a flow chart of the algorithm. The method includes two important components: the non-dominated sorting approach and the crowd comparison operator. A detailed description can be found in Ref. [41].

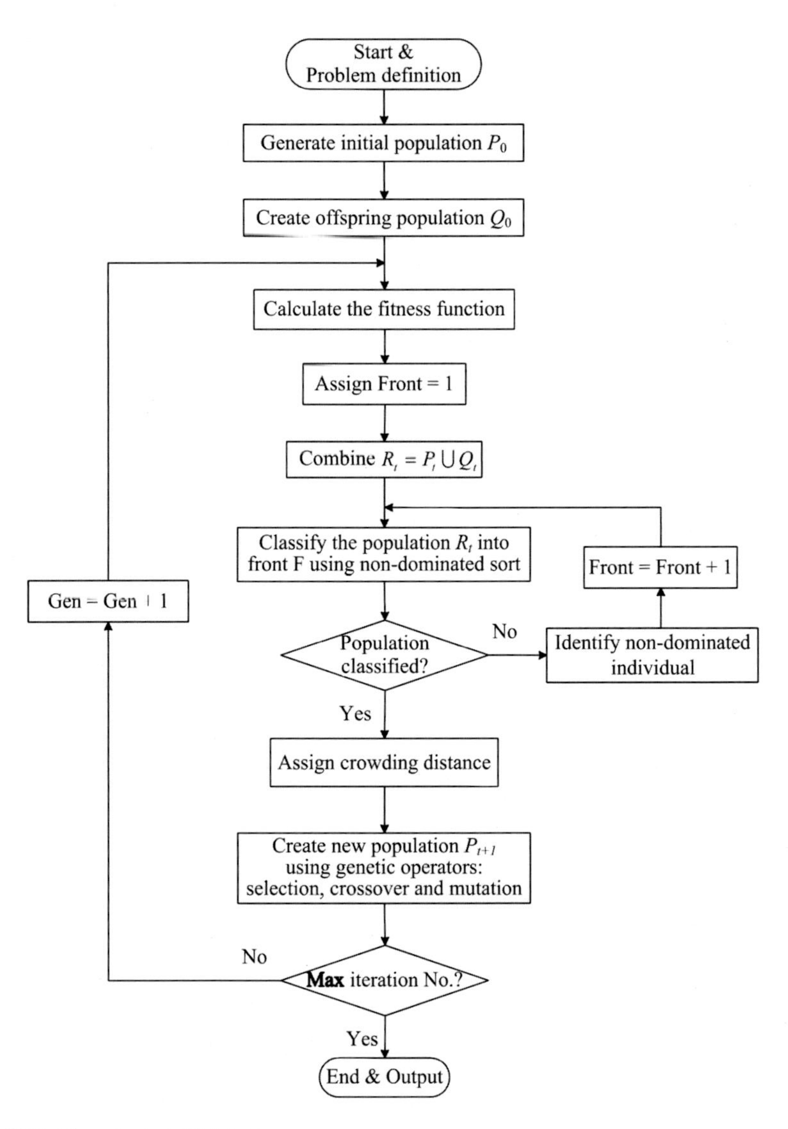

Fig. 3.11 Flowchart of NSGA II

3.3.4 MPSO

Figure 3.12 shows a general MPSO framework. MPSO has similar optimization procedures as multiobjective evolution algorithms, such as MOGA and NSGA. Many successful strategies, for example, external archive, have been introduced to MPSO. On the other hand, fitness evaluation is not a necessary step in MPSO, so that the algorithm design can be simplified. However, a global optimal position should be selected from the external archive for each particle, and this step is not required for multi-objective evolution algorithms. There are many kinds of improvements for MPSO, and many of them have been widely employed in the design optimization of electrical machines. Detail descriptions of them can be found in references [39, 40, 43].

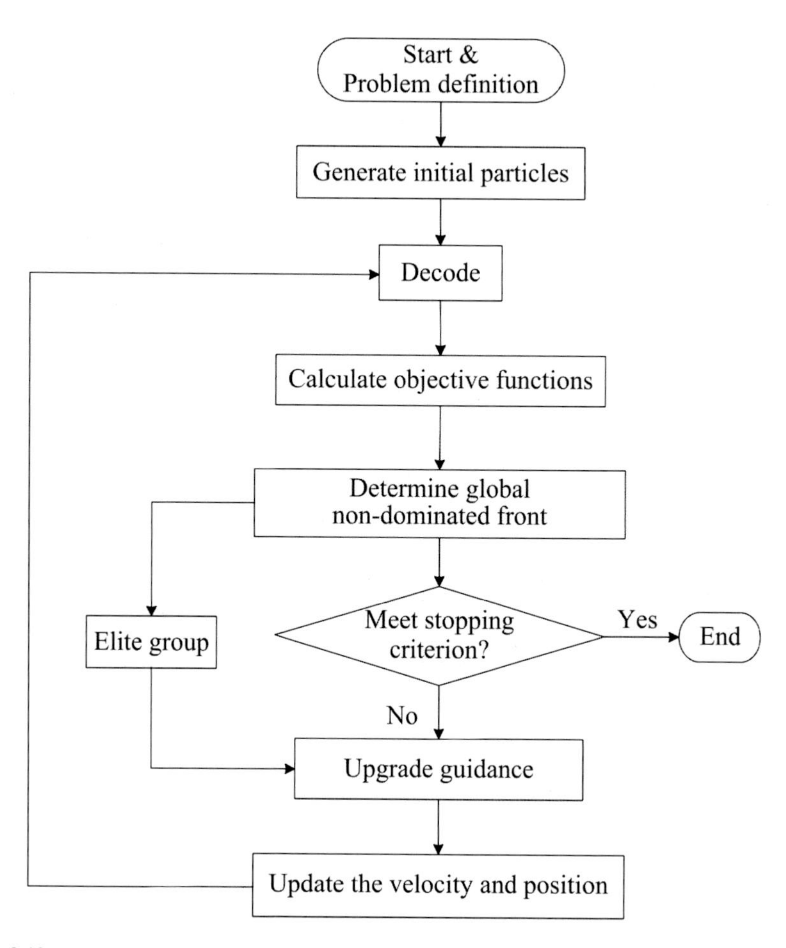

Fig. 3.12 Flowchart of MPSO algorithm

3.4 Approximate Models

3.4.1 Introduction

The above contents are about the single- and multi- optimization algorithms. As we know, they are only one important component in the optimization as well as a direct factor for the optimization efficiency. Another important issue for the optimization efficiency is related to the optimization models. FEM has been widely used in the design optimization of electrical machines. However, as mentioned previously, the computational cost of FEM is always high, especially for complex structured electrical machines, such as permanent magnetic (PM) soft magnetic composite (SMC) machines [1, 7]. As an alternative, some approximate models (or surrogate models) are employed in the practical engineering design problems to ease the computational burden of optimization process, such as RSM and Kriging model [44, 45].

Many research works have found that the optimization design based on approximate models presents an effective way to solve the aforementioned problems. By using the design of experiments (DOE) technique and statistical analysis methods, approximate models can be established as surrogate models for those physical models, such as FEMs and circuit models, so as to reduce the high simulation cost in the iterative process of optimization. Meanwhile, approximate models can degrade the nonlinearity of the practical problems, which can benefit the finding of the global optimal solution.

Generally, constructing an approximate model consists of the following two steps.

- Sampling: determine the required samples for constructing the approximate models by using the DOE technique, $\mathbf{X} = \{\mathbf{x}_1, \mathbf{x}_2, \ldots, \mathbf{x}_n | \mathbf{x}_i \in R^D\}$ and their responses $\{\mathbf{y}_1, \mathbf{y}_2, \ldots, \mathbf{y}_n | \mathbf{y}_i \in R^q\}$; and
- Modelling construction or fitting: Fit the samples (X, $\mathbf{y}$) with a suitable approximate model, and test the model accuracy with some new samples.

There are four kinds of approximate models which have been widely used in optimization design of electromagnetic devices, namely RSM, RBF model, Kriging model, and ANN model. While RSM and RBF models are parametric models, Kriging is a semi-parametric model, and ANN model a non-parametric model.

3.4.2 RSM

RSM is an empirical modeling approach for determining the relationship between various input variables and responses with various statistical criteria. It is one of the most widely used models to solve the electromagnetic optimization problems. Generally, the direct simulation can be too expensive or time consuming to carry out. RSM can effectively replace the simulation and rapidly investigate tradeoffs between various optimization tasks and conditions [46].

Generally, RSM fits the response data to lower-order quadratic polynomials using the least-square method, and the sample data are usually obtained from the uniform sampling method. Quadratic polynomials are generally used in electromagnetic problems and have been successfully developed for RSM in the form as

$$\hat{y}(x) = \beta_0 + \sum_{i=1}^{D} \beta_i x_i + \sum_{i=1}^{D} \sum_{j=1, i \le j}^{D} \beta_{ij} x_i x_j \tag{3.14}$$

where β*'s* are the regression parameters. To estimate the regression parameters, the least square method (LSM) is commonly used to minimize the quadratic sum of residual errors

$$S_e = \sum_{i=1}^{n} (y_i - \hat{y}_i)^2 \tag{3.15}$$

Let the partial derivatives for all parameters be zero, and one can obtain a linear equation system in the matrix form as the following

$$\mathbf{X}^T \mathbf{X} \boldsymbol{\beta} = \mathbf{X}^T \mathbf{y} \tag{3.16}$$

The model parameters can be calculated as

$$\boldsymbol{\beta} = (\mathbf{X}^T \mathbf{X})^{-1} \mathbf{X}^T \mathbf{y} \tag{3.17}$$

where $\mathbf{y} = [y_1, \ldots, y_n]^T$, $\boldsymbol{\beta} = [\beta_0, \beta_1, \ldots, \beta_m]^T$, and m is the dimension of parameter. For quadratic polynomials, $m = (D+1)(D+2)/2$. The structural matrix $\mathbf{X}$ is

$$\mathbf{X} = \begin{bmatrix} 1 & x_{11} & & x_{1D} & x_{11}^2 & \cdots & x_{1i}x_{1j} & \cdots & x_{1D}^2 \\ 1 & x_{21} & & x_{2D} & x_{21}^2 & \cdots & x_{2i}x_{2j} & \cdots & x_{2D}^2 \\ \vdots & \vdots & \vdots & \vdots & \vdots & \cdots & \cdots & \cdots & \cdots \\ 1 & x_{n1} & & x_{nD} & x_{n1}^2 & \cdots & x_{ni}x_{nj} & \cdots & x_{nD}^2 \end{bmatrix} \tag{3.18}$$

However, there is a conflict between the accuracy of fitting value and the step size of sample data. A smaller step size may induce higher accuracy, but the structural matrix may be poorly conditioned because the quadratic polynomials are in general globally supported in LSM. The moving least square method (MLSM) can overcome this defect. The centers of MLSM are randomly chosen and the randomness is controlled by the point density and surface geometry [47].

There are two main reasons for why the RSM based on MLSM is better than LSM. The first one is the construction of fitting functions that consist of a vector function with coefficients and a basis function instead of the traditional polynomials. Both are functions of sample points. The second is the introduction of compactly supported domain in MLSM. Within MLSM, the response of $y = f(\mathbf{x})$ is only determined by those samples in a small sub-domain around the point $\mathbf{x}$, and this sub-domain is called the compactly supported domain. Then the samples outside this sub-domain do not have any effects on the response. Therefore, a weighting function is defined in this compactly supported domain. If the weight function is constant in the whole design space, MLSM is the traditional LSM. In other words, LSM is a special case of MLSM.

In the MLSM, fitting function $\hat{y}(\mathbf{x})$ can be approximated as a sum of linearly independent function as the following

$$\hat{y}(\mathbf{x}) = \sum_{i=1}^{m} p_i(\mathbf{x}) a_i(\mathbf{x}) = \mathbf{p}^T(\mathbf{x}) \mathbf{a}(\mathbf{x}) \tag{3.19}$$

where $\mathbf{a}(\mathbf{x}) = [a_1(\mathbf{x}), a_2(\mathbf{x}), \ldots, a_m(\mathbf{x})]^T$ is the matrix of unknown coefficients, which are functions of the spatial coordinates, and m is the sample size in the fitting domain. If $\mathbf{a}(\mathbf{x})$ is constant, the LSM can be obtained. $\mathbf{p}(\mathbf{x}) = [p_1(\mathbf{x}), p_2(\mathbf{x}), \ldots, p_m(\mathbf{x})]^T$ is a matrix of complete polynomial based functions, and for a two-dimensional optimization problem, it can be expressed as

$$\mathbf{p}(\mathbf{x}) = \begin{cases} [1, x_1, x_2]^T & \text{linear base function} \\ [1, x_1, x_2, x_1^2, x_1 x_2, x_2^2]^T & \text{quadratic base function} \end{cases} \tag{3.20}$$

To calculate the coefficient vector, minimize the weighted quadratic sum of fitting errors below

$$\begin{aligned} J(\mathbf{x}) &= \sum_{i=1}^{np} \mathbf{w}(\mathbf{x} - \mathbf{x}_i)[\hat{y}(\mathbf{x}_i) - y(\mathbf{x}_i)]^2 \\ &= \sum_{i=1}^{np} \mathbf{w}(\mathbf{x} - \mathbf{x}_i)[\mathbf{p}^T(\mathbf{x}_i)\mathbf{a}(\mathbf{x}) - y(\mathbf{x}_i)]^2 \end{aligned} \tag{3.21}$$

where np is the sample size in the compactly supported domain, and $\mathbf{w}(\mathbf{x} - \mathbf{x}_i)$ a weighted function for sample $\mathbf{x}_i$. By using the LSM, the model parameter matrix can be calculated as

$$\hat{\mathbf{a}}(\mathbf{x}) = [\mathbf{p}^T(\mathbf{x})\mathbf{w}(\mathbf{x})\mathbf{p}(\mathbf{x})]^{-1}\mathbf{p}^T(\mathbf{x})\mathbf{w}(\mathbf{x})\mathbf{y}. \tag{3.22}$$

It should be noted that the fitting accuracy highly depends on the selection of weight functions, which should be equal to 0 outside the compactly supported domain. The cube spline function is widely used to get the coefficient matrix by minimizing the weighted square sum of error, which has the form as

$$W(s) = \begin{cases} 2/3 - 4s^2 + 4s^3 & s \le 1/2 \\ 4/3 - 4s + 4s^2 - 4s^3/3 & 1/2 < s \le 1 \\ 0 & s > 1 \end{cases}, \quad (3.23)$$

where $s = ||x - x_i||/s_0$, and s_0 is the radius of the compactly supported domain.

As a summary, the following lists the computational procedure of the RSM based on MLSM:

(a) Generate samples by DOE technique, and calculate the responses (objectives) of those samples;
(b) For each new sample $\mathbf{x}$ needing evaluation, implement the following steps:
   - determine the compactly supported domain for $\mathbf{x}$;
   - count the sample size inside the supported domain;
   - calculate the model coefficient vector; and
   - compute the fitting value of $\mathbf{x}$;
(c) Draw the response surface by connecting all samples.

3.4.3 RBF Model

Compared with the RSM, RBF is also an empirical modeling approach for determining the relationship between various process parameters and responses with the various desired criteria. RBF can effectively replace the time consuming simulation (or measurement) and investigate very rapidly the tradeoffs between conflicting performance criteria for optimization tasks. RBFs are commonly used in electromagnetic problems and have been successfully developed for constructing the response surface [48–50].

In general, the multivariate functions $H(\cdot) : R^d \to R$ can be efficiently evaluated if they are expressible as univariate functions $H(\cdot) = H(||\cdot||)$ of the Euclidean norm $||\cdot||$, and such functions are called RBFs. With a set of scattered points $\mathbf{x}_i (1 \le i \le n)$,the analytical expression of RBF can be given by

$$f(\mathbf{x}) = \sum_{j=1}^{n} \beta_j H\left(\left\|\mathbf{x} - \mathbf{x}_j\right\|\right) \quad (3.24)$$

where $H(r) = H\left(\left\|\mathbf{x} - \mathbf{x}_j\right\|\right)$ is the RBF, $r = \left\|\mathbf{x} - \mathbf{x}_j\right\|$ the Euclidian norm, and β_j (j = 1,2,…,n) are the unknown parameters. Three most widely used RBFs in the electromagnetic optimization problem are the Gauss, multi-quadrics (MQ) and inverse multi-quadrics (IMQ), which have the forms as

$$\text{Gauss}: H(r) = \exp(-c^2 r^2) \tag{3.25}$$

$$\text{MQ}: H(r) = (r^2 + c^2)^{1/2} \tag{3.26}$$

$$\text{IMQ}: H(r) = (r^2 + c^2)^{-1/2} \tag{3.27}$$

where $r = ||x - x_i||$, and c is the RBF constant to be determined.

As we have

$$f(\mathbf{x}_i) = y_i,\ i = 1, 2, \ldots, n \tag{3.28}$$

Then

$$\begin{cases} \beta_1 H(\|\mathbf{x}_1 - \mathbf{x}_1\|) + \beta_2 H(\|\mathbf{x}_1 - \mathbf{x}_2\|) + \cdots + \beta_n H(\|\mathbf{x}_1 - \mathbf{x}_n\|) = y_1 \\ \beta_1 H(\|\mathbf{x}_2 - \mathbf{x}_1\|) + \beta_2 H(\|\mathbf{x}_2 - \mathbf{x}_2\|) + \cdots + \beta_n H(\|\mathbf{x}_2 - \mathbf{x}_n\|) = y_2 \\ \vdots \\ \beta_1 H(\|\mathbf{x}_n - \mathbf{x}_1\|) + \beta_2 H(\|\mathbf{x}_n - \mathbf{x}_2\|) + \cdots + \beta_n H(\|\mathbf{x}_n - \mathbf{x}_n\|) = y_n \end{cases} \tag{3.29}$$

In matrix form, it is expressed as

$$\mathbf{H}\boldsymbol{\beta} = \mathbf{Y} \tag{3.30}$$

where $\mathbf{H} = \begin{bmatrix} H(\|\mathbf{x}_1 - \mathbf{x}_1\|), H(\|\mathbf{x}_1 - \mathbf{x}_2\|), \ldots, H(\|\mathbf{x}_1 - \mathbf{x}_n\|) \\ H(\|\mathbf{x}_2 - \mathbf{x}_1\|), H(\|\mathbf{x}_2 - \mathbf{x}_2\|), \ldots, H(\|\mathbf{x}_2 - \mathbf{x}_n\|) \\ \vdots \\ H(\|\mathbf{x}_n - \mathbf{x}_1\|), H(\|\mathbf{x}_n - \mathbf{x}_2\|), \ldots, H(\|\mathbf{x}_n - \mathbf{x}_n\|) \end{bmatrix}$.

When there are no superposition points, **H** is a positive definite matrix and (3.30) has an unique solution as

$$\boldsymbol{\beta} = \mathbf{H}^{-1}\mathbf{Y} \tag{3.31}$$

However, RBFs are generally globally supported and poorly conditioned (similar to the LSM). Although there are several remedies for these problems, such as domain decomposition, preconditioning, and fine tuning of the variable parameter of RBF, the compactly supported radial basis function (CSRBF) provides a promising approach.

The centers of CSRBF are randomly chosen from the points and the randomness is controlled by the point density and surface geometry. When the CSRBFs are used, the evaluation of (3.31) will not run over the whole set of n summands and the coefficient matrix will be sparse. The following two classes of CSRBF will be studied in this book [51, 52]

$$\text{CSRBF1}: H(r) = (1-r)_+^6 (6+36r+82r^2+72r^3+30r^4+5r^5), \quad (3.32)$$

and

$$\text{CSRBF2}: H(r) = (1-r)_+^8 (1+8r+25r^2+32r^3), \quad (3.33)$$

where $r = ||\mathbf{x} - \mathbf{x}_j||/r_0$, r_0 is the radius of the compactly supported domain, and $(1-r)_+$ is given by

$$(1-r)_+ = \begin{cases} 1-r & \text{if } 0 \leq r \leq 1 \\ 0 & \text{otherwise} \end{cases}. \quad (3.34)$$

3.4.4 *Kriging Model*

Given n sample points $\{\mathbf{x}_1, \mathbf{x}_2, \ldots, \mathbf{x}_n\}$ and their responses $\{y(\mathbf{x}_1), y(\mathbf{x}_2), \ldots, y(\mathbf{x}_n)\}$, for an input $\mathbf{x}$, the response value $y(\mathbf{x})$ of the Kriging model can be expressed as

$$y(\mathbf{x}) = f(\mathbf{x})^T \boldsymbol{\beta} + z(\mathbf{x}) \quad (3.35)$$

where $f(\mathbf{x})^T \boldsymbol{\beta}$ is a deterministic term for global modeling. $f(\mathbf{x})$ is a known approximation model, which is generally assumed as a polynomial and has the form of $f(\mathbf{x}) = [f_1(\mathbf{x}), f_2(\mathbf{x}), \ldots, f_q(\mathbf{x})]^T$, where q is the dimension of polynomial. $\boldsymbol{\beta}$ is the model parameter vector to be estimated. $z(\mathbf{x})$ is a random error term used for the modeling of local deviation. It is usually assumed to be a vector with the mean of zero, variance of σ^2 and covariance matrix of

$$c_{ij} = \sigma^2 \mathbf{R}[R(\mathbf{x}_i, \mathbf{x}_j)], \quad (3.36)$$

where $\mathbf{R}$ is the correlation matrix, and R the user-specified correlation function. Gaussian correlation functions are most commonly used. More details about Gaussian correlation functions and the estimation methods for the parameters in them can be found in references [53–59].

Adopting the Gaussian correlation functions, one can express the correlation matrix as

$$\mathbf{R} = \begin{bmatrix} r(\mathbf{x}_1, \mathbf{x}_1) & r(\mathbf{x}_1, \mathbf{x}_2) & \cdots & r(\mathbf{x}_1, \mathbf{x}_n) \\ r(\mathbf{x}_2, \mathbf{x}_1) & r(\mathbf{x}_2, \mathbf{x}_2) & \cdots & r(\mathbf{x}_2, \mathbf{x}_n) \\ \vdots & \vdots & \ddots & \vdots \\ r(\mathbf{x}_n, \mathbf{x}_1) & r(\mathbf{x}_n, \mathbf{x}_2) & \cdots & r(\mathbf{x}_n, \mathbf{x}_n) \end{bmatrix} \quad (3.37)$$

where

$$r(\mathbf{x}_i, \mathbf{x}_j) = \exp\{-\sum_{k=1}^{D} \alpha_k |\mathbf{x}_{ik} - \mathbf{x}_{jk}|^2\} \tag{3.38}$$

By using the best linear unbiased estimation in Statistics, the predictor of $y(\mathbf{x})$ and parameter $\boldsymbol{\beta}$ can be expressed as follows:

$$\hat{y}(\mathbf{x}) = f(\mathbf{x})^T \hat{\boldsymbol{\beta}} + r(\mathbf{x})^T \mathbf{R}^{-1}(\mathbf{y} - \mathbf{F}\hat{\boldsymbol{\beta}}) \tag{3.39}$$

$$\hat{\boldsymbol{\beta}} = (\mathbf{F}^T \mathbf{R}^{-1} \mathbf{F})^{-1} \mathbf{F}^T \mathbf{R}^{-1} \mathbf{y} \tag{3.40}$$

where $\mathbf{F}$, $r(\mathbf{x})$ and $\mathbf{y}$ are defined as

$$\mathbf{F} = \begin{bmatrix} f_1(\mathbf{x}_1) & f_2(\mathbf{x}_1) & \dots & f_q(\mathbf{x}_1) \\ \vdots & \vdots & \vdots & \vdots \\ f_1(\mathbf{x}_n) & f_2(\mathbf{x}_n) & \dots & f_q(\mathbf{x}_n) \end{bmatrix}, \tag{3.41}$$

$$r(\mathbf{x}) - \begin{bmatrix} R(\mathbf{x}, \mathbf{x}_1) \\ \vdots \\ R(\mathbf{x}, \mathbf{x}_n) \end{bmatrix}, \tag{3.42}$$

and

$$\mathbf{y} = [y(\mathbf{x}_1), y(\mathbf{x}_2), \ldots, y(\mathbf{x}_n)]^T, \tag{3.43}$$

respectively.

By using the maximum-likelihood estimation (MLE) method, the estimation of σ^2 can be obtained as

$$\hat{\sigma}^2 = \frac{1}{n}(\mathbf{y} - \mathbf{F}\hat{\boldsymbol{\beta}})^T \mathbf{R}^{-1}(\mathbf{y} - \mathbf{F}\hat{\boldsymbol{\beta}}) \tag{3.44}$$

The estimation of α_k in correlation function can be obtained from MLE. As $z(\mathbf{x})$ follows a n-dimensional normal distribution with zero 0 and covariance $\sigma^2\mathbf{R}$, the probability density function of error is

$$p(\mathbf{Y}; \boldsymbol{\beta}) = \frac{1}{\sqrt{(2\pi)^n [\det(\sigma^2 \mathbf{R})]}} \exp\left\{ \frac{(\mathbf{Y} - \mathbf{H}\boldsymbol{\beta})^T \mathbf{R}^{-1} (\mathbf{Y} - \mathbf{H}\boldsymbol{\beta})}{-2\sigma^2} \right\} \tag{3.45}$$

Substituting the predicted $\boldsymbol{\beta}$ and σ^2 , i.e. (3.40) and (3.44), into the above equation, the only unknown parameter α_k can be estimated.

In summary, implementing the Kriging method consists of estimating the parameters $\boldsymbol{\beta}$ in (3.35), σ^2 in (3.36), and the parameters α_k in Gaussian correlation functions (3.38). All the parameters can be estimated by the software package DACE (Design and Analysis of Computer Experiments) [59]. Compared with the parameter model, e.g. RSM, the Kriging model can include not only the mean trend term but also the variances of the responses. Therefore, it is superior in the modeling of local nonlinearities, and has been widely used in the optimization design of electromagnetic devices recently.

3.4.5 ANN Model

According to the model classification, the ANN model is a non-parameter model. Among the various types of network models in this research field [60–62], for data fitting and forecast, the back propagation (BP) network and RBF network may be the two most commonly used models. The BP ANN model has the form as

$$y = f(\mathbf{W}\mathbf{x} + \mathbf{b}) \tag{3.46}$$

where $\mathbf{x}$ is an input vector, y the output, f a transfer function, $\mathbf{W}$ a matrix vector of weighted value, and $\mathbf{b}$ a threshold value vector, respectively. $\mathbf{W}$ and $\mathbf{b}$ can be obtained from the model training process with the given sample points and responses.

The RBF ANN model has the form as

$$y = f(||\mathbf{W} - \mathbf{x}|| \cdot \mathbf{b}). \tag{3.47}$$

where $||\cdot||$ is the Euclid norm. Gaussian function is always used as the transfer function in this network.

3.5 Construction and Verification of Approximate Models

Figure 3.13 shows the main three design steps for the approximate models. Firstly, the samples are generated by using the DOE techniques. Then, the approximate models are constructed, including the selections of model basis functions and fitting methods. Finally, the effectiveness/accuracy of the constructed approximate models is verified [63, 64]. While the second step has been discussed previously, this section presents the first and the last steps.

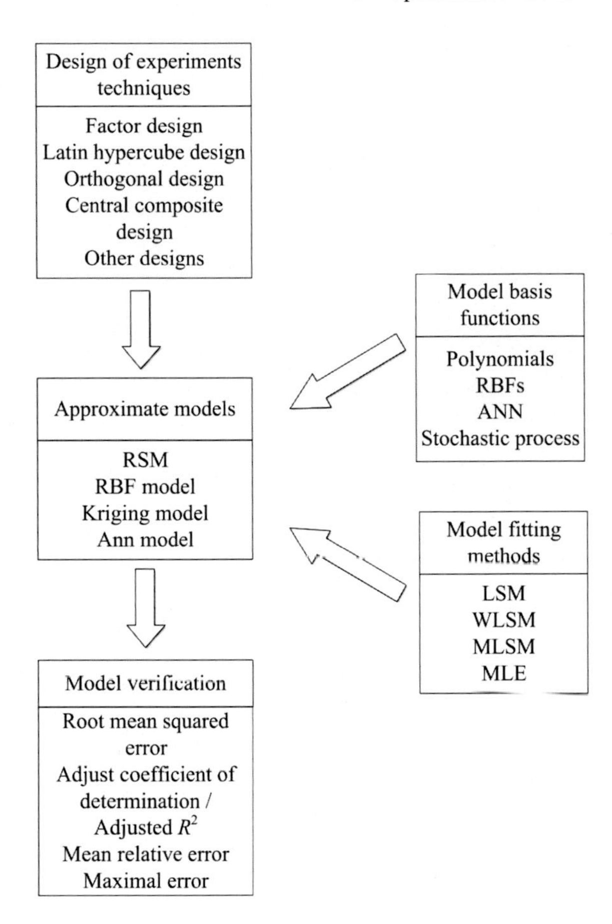

Fig. 3.13 Design flowchart of approximate models

3.5.1 DOE Techniques

Many DOE techniques have been developed, such as the full-factor design, Latin hypercube design, orthogonal design, and central composite design. To improve the fitting accuracy, the full-factor design has been employed in many situations. For example, consider an optimization problem with three parameters, and 5 levels are defined for each parameter. In total, $5^3 = 125$ samples are needed to construct an approximate model. The Latin hypercube design is a sampling method with constraints. It firstly divides the initial design space into a number of non-overlapping sub-spaces, and then sampling with equal probability is implemented in each sub-space [57, 63, 64].

3.5.2 Model Verification

After the construction of approximate models, the accuracy of effectiveness of the constructed models should be verified by some new samples. There are several verification methods as follows:

(a) Root mean square error (RMSE)
Assume that there are N_e new samples $\{\mathbf{x}_i\}$ and the true responses $\{y_i\}$, and we also have their fitting values by a kind of approximate model. The RMSE of this model is then

$$\text{RMSE} = \left(\frac{1}{N_e} \sum_{i=1}^{N_e} (y_i - \hat{y}_i)^2 \right)^{1/2} \tag{3.48}$$

The smaller the RMSE value is, the better the accuracy of fitting model.

(b) Multiple correlation coefficient
The multiple correlation coefficient, R, or the coefficient of multiple correlation is used to reveal the correlation between one factor/parameter with another one or several other factors/parameters. It is defined as

$$R = (1 - SS_e/SS_T)^{1/2} \tag{3.49}$$

where $SS_e = \sum_{i=1}^{N_e} (y_i - \hat{y}_i)^2$, $SS_T = \sum_{i=1}^{N_e} (y_i - \bar{y})^2$, and $\bar{y} = \sum_{i=1}^{N_e} \hat{y}_i$. The higher the R value is, the stronger the linear correlation is between those factors or parameters.

(c) Coefficient of determination R^2 and adjusted R^2_{adj}
The coefficient of determination is defined as the square of multiple correlation coefficient, i.e.

$$R^2 = 1 - SS_e/SS_T \tag{3.50}$$

It is a number that indicates how well the data fit a kind of model – sometimes simply a line or curve. It is one of the indexes for evaluating the modelling accuracy of approximate models. A high modelling accuracy is achieved when R^2 approaches to 1. However, this R^2 coefficient can be easily affected by the number of parameters. To avoid this undesired effect, the adjusted coefficient of determination defined as

$$R^2_{adj} = 1 - \frac{SS_e/(N_e - D - 1)}{SS_T/(N_e - 1)} \tag{3.51}$$

can be employed.

3.5.3 Modeling Examples

Two examples of classical test functions with multiple local minimums will be presented as follows to show the verifications for the RSM, RBF (Gauss type) and Kriging models.

The first one is the Rastrigin function expressed in the form of

$$f_1(x_1, x_2) = \sum_{i=1}^{2} [x_i^2 - 10\cos(2\pi x_i) + 10],\ x_i \in [-5.12, 5.12] \tag{3.52}$$

The global minimum of this function is (0, 0), and the objective is 0. There are 99 local minimums around the global minimal point (0, 0). To have a clear illustration, Fig. 3.14 shows the 3D surface profile of this function in a smaller region of $x_i \in [-2, 2]$ instead of the whole region [17, 41, 48, 63].

Figure 3.15 illustrates the RMSE curves for this Rastrigin function with the RSM, RBF, and Kriging models, where the horizontal axis is the sample size for each parameter. The minimal and maximal sample sizes are 5 and 33, respectively. For the situation of minimal sample 5, the whole optimization region [−5.12 5.12] is firstly divided into 4 parts uniformly, and then five points at [−5.12, −2.56, 0, 2.56, 5.12] are sampled with a step size of 2.56. For the situation of maximal sample 33, the whole optimization region [−5.12 5.12] is divided into 32 parts with equal length, and 33 points are sampled with a step size of 0.32. Thus, we have in total 29 situations (from 5 to 33) for each approximate model. To verify the accuracy of the constructed approximate model, 50 sample points are generated for each parameter in [−5.12, 5.12], with a step size of 0.21.

As shown in Fig. 3.15, all the three curves oscillate when the sample size is smaller than 12. When the sample size is larger than 19, the RMSE values of the RBF and Kriging models are smaller than those of the RSM, meaning that the RBF and Kriging models are better than the RSM model in terms of the modelling accuracy.

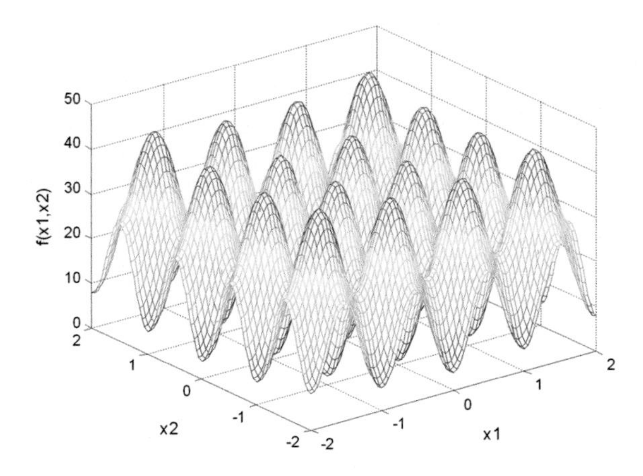

Fig. 3.14 3D surface profile of the Rastrigin function

Figure 3.16 illustrates the curves of adjusted coefficients of these approximate models for the Rastrigin function. As shown, when the sample size is larger than 19, the R_{adj}^2 values of the RBF and Kriging models approach 1 whereas those of the RSM model remain around 0.5, showing again that the RBF and Kriging models are better than the RSM model in terms of the modelling accuracy.

The second test function is defined as

$$f_2(x_1, x_2) = 0.01 \sum_{i=1}^{2} [(x_i + 0.5)^4 - 30x_i^2 - 20x_i],\ x_i \in [-5.12, 5.12] \qquad (3.53)$$

and Fig. 3.17 plots the 3D surface profile of this function.

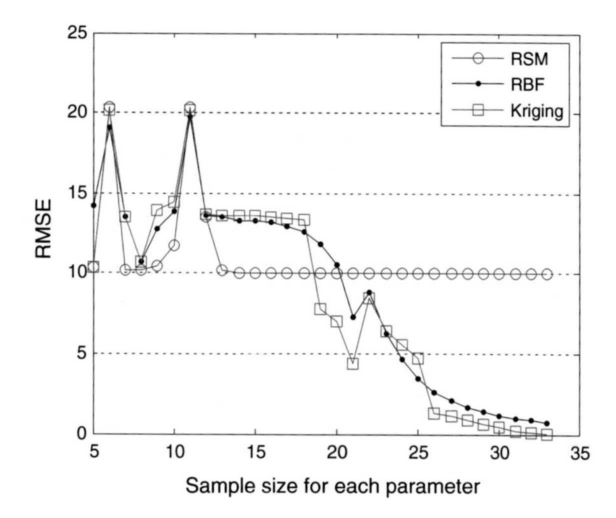

Fig. 3.15 RMSE curves of three approximate models for the Rastrigin function

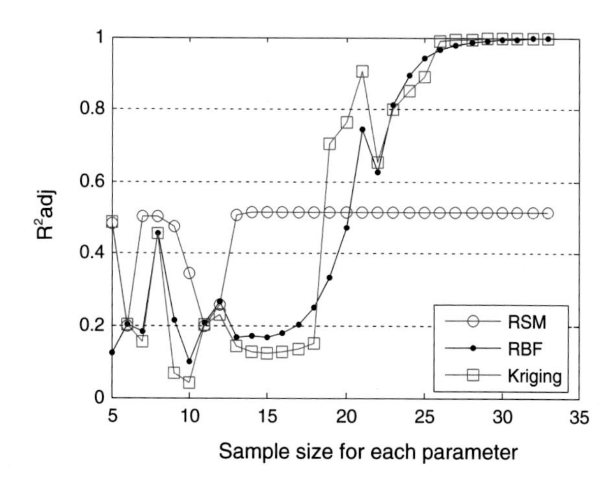

Fig. 3.16 Curves of adjusted coefficient of determination of three approximate models for the Rastrigin function

This function has four minimal points located in each of the four quadrants, respectively. The global minimum is −5.23 and is located at [−4.45, −4.45] in the third quadrant [63, 65–67]. The other minimal points are −3.68 at [3.29, 3.29] (the first quadrant), −4.46 at [−4.45, 3.29] (the second quadrant), and −4.46 at [3.29, −4.45] (the fourth quadrant).

Figure 3.18 illustrates the RMSE curves of the RSM, RBF, and Kriging models for this test function. The horizontal axis is the sample size for each parameter. Similar to the last example, the minimal and maximal sample sizes are chosen as 5 and 33, respectively. To verify the accuracy of the constructed approximate models, 100 sample points are generated for each parameter. As shown, the RMSE values of the RBF and Kriging models are significantly smaller than those of the RSM model. It can also be seen that the Kriging model has fastest convergence and the smallest RMSE values for almost all sample sizes.

Figure 3.19 illustrates the curves of the adjusted coefficient of these approximate models for the second test function. As shown, when the sample size is larger than

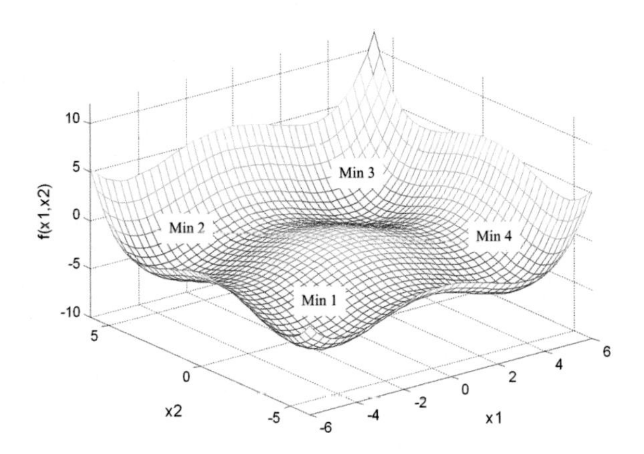

Fig. 3.17 3D surface profile of the second test function

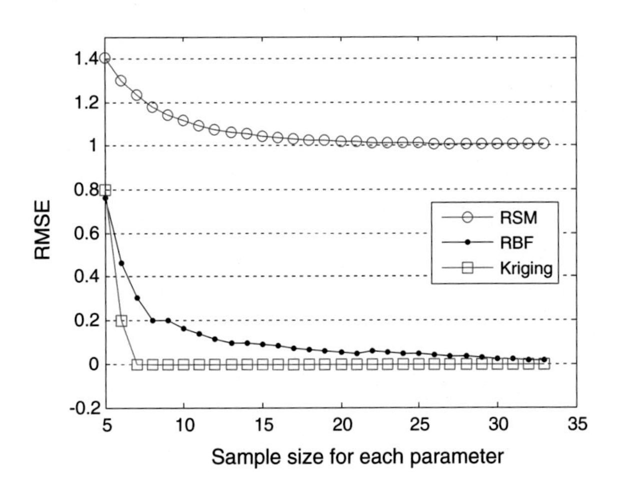

Fig. 3.18 RMSE curves of three approximate models for the second test function

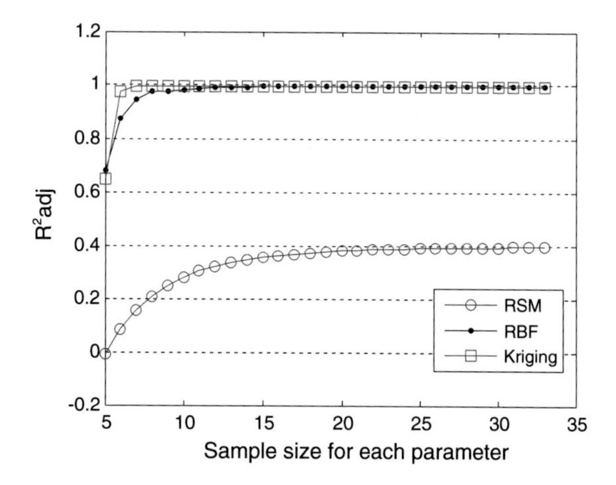

Fig. 3.19 Curves of adjusted coefficient of determination of three approximate models for the second test function

19, the R_{adj}^2 values of the RBF and Kriging models approach 1 quickly, whereas those of the RSM model remain around 0.4, which means the RBF and Kriging models are better than the RSM model in terms of the modelling accuracy.

As shown by these two examples, the RBF and Kriging models are better than the RSM model in terms of the modelling accuracy, and in both cases, the Kriging model is better than the RBF model. It can also be seen that the modelling accuracy of these models for the second test function is higher than that for the first one because of the stronger nonlinearity and much more number of minimums of the first function.

3.6 Summary

This chapter presents a brief summary of the most commonly used numerical optimization algorithms for electrical machines and drive systems, including the classic gradient-based algorithms, modern intelligent algorithms, and multi-objective optimization algorithms. Four kinds of approximate/surrogate models, namely RSM, RBF, Kriging, and ANN models, have been presented with details. Two examples are studied to illustrate the procedure of construction and verification of approximate models.

References

1. Lei G, Wang TS, Guo YG, Zhu JG, Wang SH (2014) System level design optimization method for electrical drive system: deterministic approach. IEEE Trans Ind Electron 61 (12):6591–6602
2. Yao D, Ionel DM (2013) A review of recent developments in electrical machine design optimization methods with a permanent-magnet synchronous motor benchmark study. IEEE Trans Ind Appl 49(3):1268–1275

3. Chechurin VL, Korovkin NV, Hayakawa M (2007) Inverse problems in electric circuits and electromagnetics. Springer, New York
4. Coelho L, Alotto P (2007) Electromagnetic device optimization by hybrid evolution strategy approaches. Int J Comput Math Electr Electron Eng (COMPEL) 26(2):269–279
5. Nguyen TD, Lanfranchi V, Doc C, Vilain JP (2009) Comparison of optimization algorithms for the design of a brushless DC electric machine with travel time minimization. In: Proceedings of the 8th international symposium on advanced electromechanical motion systems & electric drives (ELECTROMOTION), pp 1–6
6. Di Barba P (2010) Multiobjective shape design in electricity and magnetism. Lect Notes Elect Eng 47
7. Lei G, Shao KR, Guo YG, Zhu JG (2012) Multiobjective sequential optimization method for the design of industrial electromagnetic devices. IEEE Trans Magn 48(11):4538–4541
8. Reyes-Sierra M, Coello CAC (2006) Multi-objective particle swarm optimizers: A survey of the state-of-the-art. Int J Comput Intell Res 2(3):287–308
9. Dubas F, Sari A, Hissel D, Espanet C (2008) A comparison between CG and PSO algorithms for the design of a PM motor for fuel cell ancillaries. In: Proceedings of the IEEE Vehicle Power and Propulsion Conference, pp 1–7
10. Razik H, Defranoux C, Rezzoug A (2000) Identification of induction motor using a genetic algorithm and a quasi-Newton algorithm. In: Proceedings of the IEEE International Power Electronics Congress, pp 65–70
11. Qteish A, Hamdan M (2010) Hybrid particle swarm and conjugate gradient optimization algorithm. Proc Adv Swarm Intell 6145:582–588
12. Popa D-C, Micu DD, Miron O-R, Szabo L (2013) Optimized design of a novel modular tubular transverse flux reluctance machine. IEEE Trans Magn 49(11):5533–5542
13. Whitley D (1994) A genetic algorithm tutorial. Stat Comput 4(2):65–85
14. Hasanien HM, Abd-Rabou AS, Sakr SM (2010) Design optimization of transverse flux linear motor for weight reduction and performance improvement using response surface methodology and genetic algorithms. IEEE Trans Energy Convers 25(3):598–605
15. Hasanien HM (2011) Particle swarm design optimization of transverse flux linear motor for weight reduction and improvement of thrust force. IEEE Trans Ind Electron 58(9):4048–4056
16. Chun JS, Kim MK, Jung HK (1997) Shape optimization of electromagnetic devices using immune algorithm. IEEE Trans Magn 33(2):1876–1879
17. Campelo F, Guimaraes FG, Igarashi H et al (2005) A clonal selection algorithm for optimization in electromagnetics. IEEE Trans Magn 41(5):1736–1739
18. Ho SL, Yang SY, Ni G et al (2005) A particle swarm optimization-based method for multiobjective design optimizations. IEEE Trans Magn 41(5):1756–1759
19. http://www.obitko.com/tutorials/genetic-algorithms/ga-basic-description.php
20. Srinivas M (1994) Genetic algorithms: a survey. Computer 6:17–26
21. Storn R, Price K (1997) Differential evolution - a simple and efficient heuristic for global optimization over continuous spaces. J Global Optim 11:341–359
22. Lei G, Shao KR, Guo Y, Zhu J, Lavers JD (2008) Sequential optimization method for the design of electromagnetic device. IEEE Trans Magn 44(11):3217–3220
23. Lei G, Shao KR, Guo Y, Zhu J, Lavers JD (2009) Improved SOM for high dimensional electromagnetic optimization problems. IEEE Trans Magn 45(10):3993–3996
24. Larranaga P, Lozano JA (2002) Esitimation of distribution algorithm: a new tool for evolutionary computation. Kluwer Acadmic Publisher, Boston
25. Pelikan M (2005) Hierarchical Bayesian optimization algorithm: toward a new generation of evolutionary algorithms. Springer, New York
26. Hauschild M, Pelikan M (2011) An introduction and survey of estimation of distribution algorithms. Swarm and Evol Comput 1(3):111–128
27. Campelo F, Guimaraes FG, Ramirez JA, Igarashi H (2009) Hybrid estimation of distribution algorithm using local function approximations. IEEE Trans Magn 45:1558–1561

28. Ho SL, Yang SY, Ni GZ, Wong HC (2006) A particle swarm optimization method with enhanced global search ability for design optimizations of electromagnetic devices. IEEE Trans Magn 42(4):1107–1110
29. Al-Aawar N, Hijazi TM, Arkadan AA (2011) Particle swarm optimization of coupled electromechanical systems. IEEE Trans Magn 47(5):1314–1317
30. Elhossini A, Areibi S, Dony R (2010) Strength Pareto particle swarm optimization and hybrid EA-PSO for multi-objective optimization. Evol Comput 18(1):127–156
31. Beniakar ME, Sarigiannidis AG, Kakosimos PE, Kladas AG (2014) Multiobjective evolutionary optimization of a surface mounted PM actuator with fractional slot winding for aerospace applications. IEEE Trans Magn 50(2), Article 7016404
32. dos Santos Coelho L, Barbosa LZ, Lebensztajn L (2010) Multiobjective particle swarm approach for the design of a brushless DC wheel motor. IEEE Trans Magn 46(8):2994–2997
33. Alotto P, Baumgartner U, Freschi F, Köstinger A, Magele Ch, Renhart W, Repetto M (2008) SMES optimization benchmark extended: introducing Pareto optimal solutions into TEAM22. IEEE Trans Magn 44(6):1066–1069
34. Guimaraes FG, Campelo F, Saldanha RR, Igarashi H, Takahashi RHC, Ramirez JA (2006) A multiobjective proposal for the TEAM benchmark problem 22. IEEE Trans Magn 42(4):1471–1474
35. Lebensztajn L, Coulomb JL (2004) TEAM workshop problem 25: a multiobjective analysis. IEEE Trans Magn 40(2):1402–1405
36. Dias AHF, Vasconcelos JA (2002) Multiobjective genetic algorithms applied to solve optimization problems. IEEE Trans Magn 38(2):1133–1136
37. dos Santos Coelho L, Alotto P (2008) Multiobjective electromagnetic optimization based on a nondominated sorting genetic approach with a chaotic crossover operator. IEEE Trans Magn 44(6):1078–1081
38. Caravaggi Tenaglia G, Lebensztajn L (2014) A multiobjective approach of differential evolution optimization applied to electromagnetic problems. IEEE Trans Magn 50(2), Article 7015404
39. Ho SL, Yang JQ, Yang SY, Bai YN (2015) Integration of directed searches in particle swarm optimization for multiobjective optimization. IEEE Trans Magn 51(3) Article: 7000804
40. Al-Aawar N, Hijazi TM, Arkadan AA (2011) Particle swarm optimization of coupled electromechanical systems. IEEE Trans Magn 47(5):1314–1317
41. Deb K, Pratap A, Agarwal S, Meyarivan T (2002) A fast and elitist multi-objective genetic algorithm: NSGA-II. IEEE Trans Evol Comput 6(2):182–197
42. Srinivas N, Deb K (1994) Multiobjective optimization using nondominated sorting in genetic algorithms. Evol Comput 2(3):221–248
43. Sooksaksun N (2012) Pareto-based multi-objective optimization for two-block class-based storage warehouse design. Ind Eng Manage Syst 11(4):331–338
44. Wang LD, Lowther DA (2006) Selection of approximation models for electromagnetic device optimization. IEEE Trans Magn 42(2):1227–1230
45. Mendes MHS, Soares GH, Coulomb J-L, Vasconcelos JA (2013) Appraisal of surrogate modeling techniques: a case study of electromagnetic device. IEEE Trans Magn 49(5):1993–1996
46. Dyck D, Lowther DA, Malik Z et al (1998) Response surface models of electromagnetic devices and their application to design. IEEE Trans Magn 35(3):1821–1824
47. Liu WK, Li S, Belytschko T (1997) Moving least-square reproducing kernel methods (I) Methodology and convergence. Comput Methods Appl Mech Eng 143:113–154
48. Ishikawa T, Tsukui Y, Matsunami M (1999) A combined method for the global optimization using radial basis function and deterministic approach. IEEE Trans Magn 35(3):1730–1733
49. Gutmann HM (2001) A radial basis function method for global optimization. J Global Opt 19 (3):201–227
50. Lei G, Yang GY, Shao KR, Guo YG, Zhu JG, Lavers JD (2010) Electromagnetic device design based on RBF models and two new sequential optimization strategies. IEEE Trans Magn 46(8):3181–3184

51. Wu ZM (1995) Compactly supported positive definite radial basis functions. Adv Comput Math 4:283–292
52. Song KZ, Zhang X, Lu MW (2004) Meshless method based on collocation with consistent compactly supported radial basis functions. Acta Mech Sin 20(5):551–557
53. Lebensztajn L, Marretto CAR, Costa MC et al (2004) Kriging: a useful tool for electromagnetic device optimization. IEEE Trans Magn 40(2):1196–1199
54. Xia B, Pham M-T, Zhang YL, Koh C-S (2013) A global optimization algorithm for electromagnetic devices by combining adaptive Taylor Kriging and particle swarm optimization. IEEE Trans Magn 49(5):2061–2064
55. Lim D-K, Woo D-K, Kim I-W, Ro J-S, Jung H-K (2013) Cogging torque minimization of a dual-type axial-flux permanent magnet motor using a novel optimization algorithm. IEEE Trans Magn 49(9):5106–5111
56. Kim JB, Hwang KY, Kwon BI (2013) Optimization of two-phase in-wheel IPMSM for wide speed range by using the Kriging model based on Latin hypercube sampling. IEEE Trans Magn 47(5):1078–1081
57. Lei G, Guo YG, Zhu JG, Wang TS, Chen XM, Shao KR (2012) System level six sigma robust optimization of a drive system with PM transverse flux machine. IEEE Trans Magn 48 (2):923–926
58. Lei G, Guo YG, Zhu JG, Chen XM, Xu W (2012) Sequential subspace optimization method for electromagnetic devices design with orthogonal design technique. IEEE Trans Magn 48 (2):479–482
59. Lophaven SN, Nielsen HB, Sondergaard J (2002) DACE: A MATLAB Kriging toolbox version 2.0. Technical Report IMM-TR-2002–12. Technical University of Denmark, Copenhagen
60. Dorica M, Giannacopoulos DD (2006) Response clustering for electromagnetic modeling and optimization. IEEE Trans Magn 42(4):1127–1130
61. Marinova I, Panchev C, Katsakos D (2000) A neural network inversion approach to electromagnetic device design. IEEE Trans Magn 36(4):1080–1084
62. Burrascano P, Fiori S, Mongiardo M (1999) Review of artificial neural networks applications in microwave computer-aided design. Int J RF Microw CAE 9(3):158–174
63. Lei G (2009) Statistical analysis method for electromagnetic inverse problems . Ph D thesis. Huazhong University of Science and Technology, Wuhan, China (in Chinese)
64. Wu CFJ, Hamada MS (2000) Experiments: planning, analysis and parameter design optimization. Wiley, New York
65. Rashid K, Farina M, Ramirez JA et al (2001) A comparison of two generalized response surface methods for optimization in electromagnetics. Int J Comput Math Electr Electron Eng (COMPEL) 20(3):740–752
66. Coulomb JL, Kobetski A, Caldora M, Costa et al (2003) Comparison of radial basis function approximation techniques. Int J Comput Math Electr Electron Eng (COMPEL) 22(3):616–629
67. Pahner U, Hameyer K (2000) Adaptive coupling of differential evolution and multiquadrics approximation for the tuning of the optimization process. IEEE Trans Magn 36(4):1047–1050

Chapter 4
Design Optimization Methods for Electrical Machines

Abstract This chapter presents the design optimization methods for electrical machines in terms of different optimization situations, including low- and high-dimensional, single- and multi- objectives and disciplines. Firstly, the traditional design optimization methods are briefly reviewed, and the challenges presented. Then, five new types of design optimization methods are presented to improve the optimization efficiency of electrical machines, particularly those complex structured permanent magnet machines, in terms of different optimization situations. They are (a) a sequential optimization method for design optimization of low-dimensional problems of electromagnetic devices including electrical machines, (b) a multi-objective sequential optimization method for engineering multi-objective problems, (c) a multi-level design optimization method (or sequential subspace optimization method) for high dimensional problems, (d) a multi-level genetic algorithm for high dimensional optimization problems as well, and (e) the multi-disciplinary design optimization method. Design examples with detailed experimental and optimization results are illustrated for each optimization method.

Keywords Design optimization · Sequential optimization method · Multi-objective sequential optimization method · Multi-level optimization method · Multi-level genetic algorithm · Multi-disciplinary optimization method · Permanent magnet motors

4.1 Introduction

Design optimization methods actually consist of two parts, design methods with analysis models, and optimization methods with algorithms. In Chap. 2, the popular design methods and analysis models for electrical machines and drive systems are reviewed. As shown, the design of electrical machines is a complex multi-disciplinary or multi-physics problem, including electromagnetics, thermotics, mechanics and control, and each discipline has its own design methods and analysis models [1–3]. For example, the electromagnetic design is mainly based on analytical

G. Lei et al., *Multidisciplinary Design Optimization Methods for Electrical Machines and Drive Systems*, Power Systems,
DOI 10.1007/978-3-662-49271-0_4

model, magnetic circuit model and finite element model (FEM). The thermal design is mainly based on FEM and thermal network model. The control system design is mainly based on the topologies of power electronic circuits and control algorithms, such as the field oriented control algorithm and the direct torque control algorithm.

In Chap. 3, the popular optimization algorithms and approximate models used in optimization of electrical machines as well as other electromagnetic devices are discussed. The optimization algorithms include the classical gradient-based algorithms and modern intelligent algorithms, such as genetic algorithms (GA), differential evolution algorithm (DEA), and multi-objective genetic algorithms (MOGA). The approximate models include the response surface model (RSM), radial basis function (RBF) model, compactly supported radial basis function (CSRBF) model, and Kriging model [4–6].

The general procedures for design optimization of electrical machines are listed as follows:

(1) Determine the design analysis model for the investigated machine, including selection of material, motor type and topology, and develop the multi-disciplinary analysis model, for example, electromagnetic-thermal coupled model based on FEM.
(2) Establish the optimization model, including the definition of objectives (such as maximization of output power and efficiency, and minimization of cost), constraints (such as volume, mass and temperature rises) and design parameters (such as material and structural parameters). The optimization regions of these parameters and their types, such as discrete and continuous, have to be defined in this step as well.
(3) Select an optimization method and optimize the established optimization model. There are many kinds of available optimization methods. The most popular one is the direct optimization method, which uses an optimization algorithm to optimize the optimization model established on FEM.
(4) Validate the effectiveness of the obtained optimal solutions by experiments or other ways. If necessary, update the optimization model in terms of the experimental results and do the optimization again.

4.2 Classical Optimization Methods

In general, there are two kinds of classical optimization methods, and they are

(1) The direct optimization methods, and
(2) The optimization methods based on approximate models.

The direct optimization methods use the optimization algorithms to directly optimize the physical models, such as the analytical model, magnetic circuit model, and FEM for the design of electrical machines. For example, the conjugate gradient algorithm and sequential quadratic programing algorithm have been employed to optimize several kinds of motors based on the analytical models for electromagnetic analysis. The most popular formula of this type of methods is the combination of intelligent algorithms and FEM, such as GA&FEM, DEA&FEM and MOGA&FEM [2, 7–11].

This approach can present global optimal design schemes for electrical machines. However, the computational cost of this kind of optimization methods is always huge due to the extensive computational burdens of FEM, practically for some complex structured electrical machines requiring 3D FEM, such as the transverse flux machines (TFM) and claw pole motors. Most importantly, the computational cost will increase greatly with the increase of problem dimension. For example, for a motor design problem with four parameters, about 4,000 ($4 \times 5 \times 200$, where 4×5 is the population size and 200 the general iteration number of GAs) FEM samples are needed if the GA&FEM method is applied. However, if a motor has 10 design parameters, about 20,000 ($10 \times 5 \times 200$) FEM samples are required, which is a huge computational burden for many situations.

An effective way to solve this problem is to use the second kind of optimization method: the optimization based on approximate models. This method replaces the FEM with a kind of approximate model, such as RSM and Kriging, so as to form the optimization method, GA&RSM and GA&Kriging. As mentioned in Chap. 3, the approximate models can degrade the nonlinearity of the optimization problem. Therefore, the optimization efficiency can be improved significantly.

This method is relatively simple to implement. However, the optimization accuracy is an important problem for this kind of methods, particularly for high dimensional problems. Actually, it is very hard or impossible to replace the FEM with approximate models for high dimensional problems, because they cannot approximate high dimensional problems with sufficient accuracy by using limited number of samples. As we know, the first step in the construction of approximation models is to obtain the initial samples by using the design of experiments (DOE) technique. If 5 samples are required for each parameter in a design problem with ten parameters, 5^{10} FEM samples are required in total, which is greater than that required by direct optimization method of GA and FEM [1].

Therefore, the classical or traditional direct optimization methods based on both FEM and the approximation models have challenges for the design optimization of electrical machines, especially for the high dimensional design optimization problems. The following sections will present several new optimization methods.

4.3 Sequential Optimization Method

4.3.1 Method Description

In the traditional optimization methods, the optimization models and intelligent algorithms are processed almost separately though the optimization is indeed a

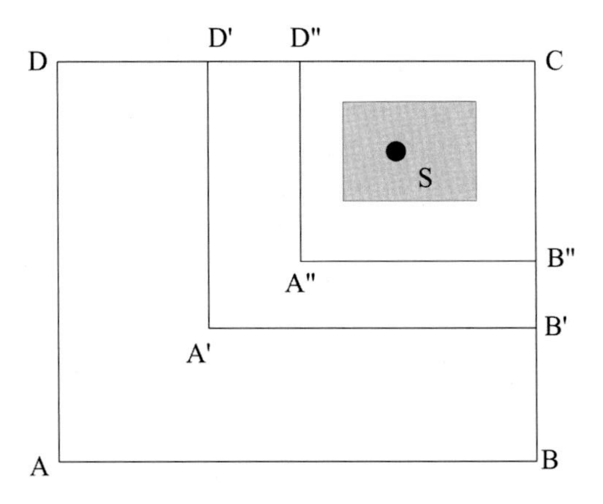

Fig. 4.1 Optimization strategy for SOM

simultaneous updating process about them. This is one of the main reasons why the traditional optimization methods cause huge computational costs. To overcome this problem, a sequential optimization method (SOM) was developed to reduce the computational cost for single-objective low-dimensional optimization problems of electromagnetic devices in 2008 [12, 13].

Figure 4.1 illustrates the optimization strategy of SOM. Assume that square ABCD is the initial design space and point S is the optimal point. The traditional direct optimization method, e.g. GA&FEM, searches the whole design space for the optimal point by an iterative process, and many new samples in the population are required in the whole design space. However, the optimal point is located in a small subspace (shown as the shaded rectangle) around point S. If we can find this interested subspace and sample more points in it instead of the whole design space, the optimization efficiency would be improved greatly. Based on this idea, the main question is how to find this subspace efficiently. SOM is a method to deal with this problem. It can reduce the design space step by step. As shown in Fig. 4.1, it reduces the design space from ABCD to A′B′CD′ in the first step, then to A″B″CD″ in the second step, and so on until the shaded rectangle is reached in the last step.

Figure 4.2 depicts a brief flowchart of SOM. Basically, SOM can be regarded as a space-to-space optimization strategy compared with the point-to-point optimization strategy of the traditional intelligent algorithms, such as GA and DEA. SOM consists of two optimization processes, coarse and fine optimization processes. The main aim of the former is to reduce the design space to a small space (the shaded rectangle as shown in Fig. 4.1). The purpose of the latter is to update the model in the local space and find the optimal solutions [12, 13].

SOM is conducted in the following six main steps:

Step 1: Define optimization model, such as objectives, constraints and design parameters. Select the approximate models and optimization algorithms that will be used in the SOM, and define the algorithm parameters, such as population, genetic operators and maximal iteration number.

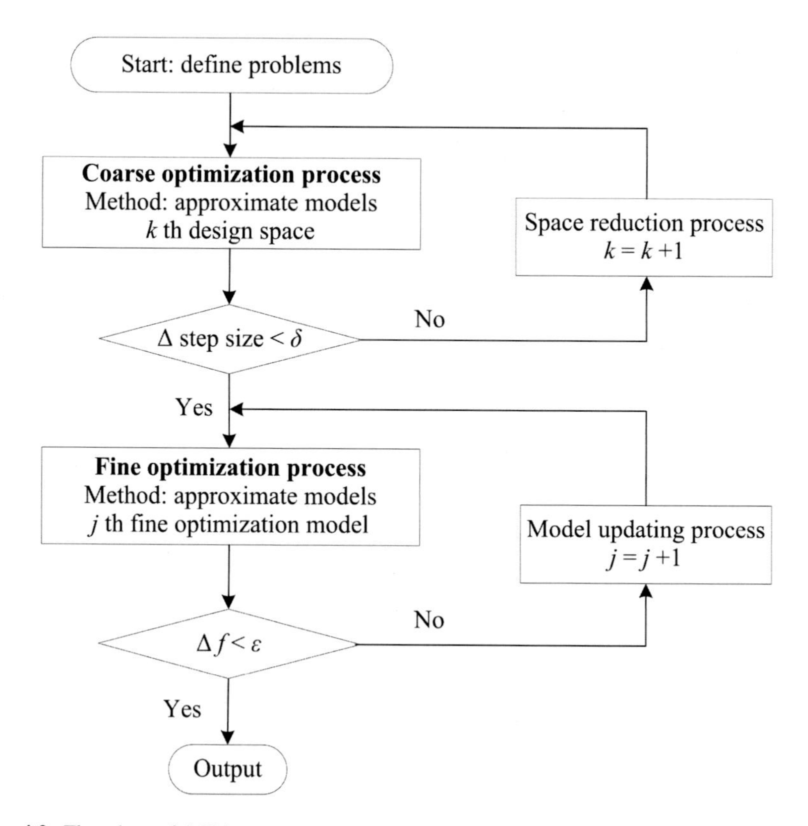

Fig. 4.2 Flowchart of SOM

Step 2: Use the coarse optimization process (COP) to reduce the design space. The traditional optimization methods based on approximate models will be employed in this step. The DOE is required in this method to generate samples for the construction of approximate models.

In general, one can assume that the design space of the *k*th COP is $[x_{li}^{(k)}, x_{ui}^{(k)}]$, $i = 1, 2, \ldots, D$, where D is the dimension, $l_i^{(k)}$ the interval of the *i*th variable, $h_i^{(k)} = h_i^{(k-1)}/2$ the step size, and $N_i^{(k)}$ the sample size, respectively. The sample set $\mathrm{S}^{(k)}$ can be obtained by using a DOE technique, e.g. the full-factor design. Based on these samples, the approximate model can be constructed. Finally, through the optimization of the model, the current optimal point $\mathbf{x}_o^{(k)} = \{x_{oi}^{(k)} | i = 1, 2, \ldots, D\}$ and objective $f^{(k)}$ can be obtained.

Note that the step size must ensure that the minimum number of sample points is no less than 3. Otherwise, a singular matrix may appear in the matrix inversion process of model construction.

Step 3: Terminate COP by the step size. If $h_i^{(k)}/l_i^{(1)} < \delta$, where δ is a positive constant and can be a value in [1 %, 5 %], stop COP and go to step 5. Otherwise, go to step 4.

Step 4: Reduce the current design space with the obtained optimal value. Under the boundary condition of design space, the design space of next step can be updated as follows:

$$\begin{cases} x_{li}^{(k+1)} = \max\left\{x_{li}^{(k)}, round\left(\frac{x_{oi}^{(k)} - \Delta l}{\Delta h}\right)\Delta h\right\} \\ x_{ui}^{(k+1)} = \min\left\{x_{ui}^{(k)}, round\left(\frac{x_{oi}^{(k)} + \Delta l}{\Delta h}\right)\Delta h\right\} \end{cases}, \tag{4.1}$$

where function *round*(x) rounds x to the nearest integer, $\Delta l = l_i^{(k)}/nl$, $\Delta h = h_i^{(k)}/nh$, and nl and nh are the reduction factors.

For a practical problem, three parameters, nl, nh and N, can be used to determine the construction of an approximate model. nl can be 4, 6 or 8, meaning that the corresponding intervals of reduced space are 1/2, 1/3 and 1/4 of the former space, respectively. nh can be 2, 4, or 8, which let function *round*(x) round x with 1/2, 1/4 and 1/8 of the current step size, respectively, and N can be 3, 4 or 5 for a standardization space of [0, 1].

To find the best values of these parameters, the Monte Carlo analysis (MCA) method is employed. Firstly, assume that the initial design space is [0, 1] and nl is 4. Then, generate 10^6 random numbers by using the Monte Carlo method as the optimal results of an approximate model. Thereafter, use these numbers to reduce the design space under $N = \{3, 4, 5\}$ and $nh = \{2, 4, 8\}$. The target of this analysis is to compare the mean of the errors between the current optimal results and the average of the reduced space $(x_{li}^{(k)} + x_{ui}^{(k)})/2$. Table 4.1 lists the MCA results for the case of $nl = 4$. Similarly, Tables 4.2 and 4.3 list the results for the cases of $nl = 6$ and $nl = 8$, respectively.

As shown, the mean error decreases with the increase of N and nh for all three cases. Thus, in the later implementation, the default value of nh is 8 and N is 5. The

Table 4.1 Mean of errors for space reduction strategy with $nl = 4$

N	$nh = 2$	$nh = 4$	$nh = 8$
3	0.0625	0.0469	0.0390
4	0.0555	0.0416	0.0364
5	0.0469	0.0391	0.0352

Table 4.2 Mean of errors for space reduction strategy with $nl = 6$

N	$nh = 2$	$nh = 4$	$nh = 8$
3	0.0417	0.0260	0.0200
4	0.0417	0.0278	0.0209
5	0.0260	0.0199	0.0168

Table 4.3 Mean of errors for space reduction strategy with $nl = 8$

N	$nh = 2$	$nh = 4$	$nh = 8$
3	0.0625	0.0313	0.0195
4	0.0243	0.0243	0.0156
5	0.0313	0.0195	0.0137

factor nl must be selected in terms of the practical problems. If we have some experience about the problem or the given space is very large and we want to reduce the design space quickly, $nl = 6$ or 8 can be selected. Otherwise, $nl = 4$ may be a better choice. In the later discussions, the default value of nl is 4 if there is no further explanation.

Secondly, to illustrate the efficiency of the new method, two examples are shown below. For the first example, assume that the initialization of design space is [0, 1] and the step size is 0.2; which means that six sample points are composed of the first sample data $S^{(1)} = \{0.0, 0.2, 0.4, 0.6, 0.8, 1.0\}$. If the optimal value is 0.35, the next sample space is [0.1, 0.6], and the new sample data are $S^{(2)} = \{0.1, 0.2, 0.3, 0.4, 0.5, 0.6\}$, i.e. three points have been sampled in the last step. In other words, the computational cost is reduced by 50 %. Similarly, if the optimal value is 0.15, the new sample space is [0, 0.4], and new sample data are $S^{(2)} = \{0, 0.1, 0.2, 0.3, 0.4\}$. In this case, the computational cost is reduced by 60 %.

The second example is about the value of nh. If nh is 2 and the current optimal value is 0.5, the next sample space is [0.5, 1.0]. If nh is 8, the next sample space is [0.25, 0.75]. Therefore, we can minimize the distance between the optimal results and the average of the next design space by using a bigger nh.

Step 5: Use fine optimization process (FOP) to find the final optimal results. To ensure the accuracy and robustness of optimization process, the local multipoint sample updating method is proposed here. Given the current optimal value, the next sample set $S^{(k+1)}$ is updated by

$$S^{(k+1)} = S_{(k)} \cup \{\mathbf{x}_o^{(k)} \pm \Delta\mathbf{x}_o^{(p)} | p = 1, \ldots, N_p\}, \tag{4.2}$$

which is constructed by N_p perturbations around the current optimal value, where N_p is the number of new samples, and can be defined as 2^D, meaning that two new points are sampled for each variable.

Step 6: Terminate the optimization process according to predefined error. If $|\Delta f^{(k)}/f^{(k)}| < \varepsilon$, where ε is a small positive constant, and can be a value in [1, 5 %], stop the optimization process and output the final optimal results; otherwise go to step 5 until the termination condition is satisfied.

4.3.2 Test Example 1—A Mathematical Test Function

A test function shown in Chap. 3 is used here to verify the efficiency of the proposed SOM [12], which has the form as

$$f_2(x_1, x_2) = 0.01 \sum_{i=1}^{2} [(x_i + 0.5)^4 - 30x_i^2 - 20x_i],\ x_i \in [-5.12, 5.12] \tag{4.3}$$

The parameters in SOM are $\delta = \varepsilon = 5\ \%$ and $N^{(1)} = [5, 5]$. The DEA is used as the optimization algorithm in this example [14]. The algorithm parameters are chosen as 0.8 for the mutation scaling factor, 0.8 for the crossover factor, 1000 for the maximum number of iteration, and 100 for the maximum stall generation, which is selected for the stop criterion [12, 13].

Table 4.4 tabulates the optimization results obtained by SOM with five different approximate models. They are the general RSM model based on least square method, improved RSM model based on moving least square (MLS) method, Gauss RBF model, CSRBF 1 model, and Kriging model. As shown, the obtained optimal solutions by using the SOM are of good accuracy compared with the exact values.

4.3.3 Test Example 2—Superconducting Magnetic Energy Storage

Superconducting magnetic energy storage (SMES) is an attractive research area in application of superconducting materials. Taking advantage of the property of low power loss and fast response of superconducting magnets, SMES can be employed as a multi-functional electromagnetic system to store and release electricity for power systems with the connection of power electronic converters. SMES is able to store large amount of energy with very low power losses, so as to improve the power supply quality and enhance the stability and reliability of power systems.

TEAM Workshop Problem 22 deals with the optimization design problem of a specific SMES, which is often used as a benchmark problem to verify and compare

Table 4.4 Optimization results of SOM for an analytic function

Method	x_1	x_2	f
Exact	−4.4538	−4.4538	−5.2328
RSM	−4.4060	−4.4060	−5.2299
RSM (MLS)	−4.4060	−4.4060	−5.2299
RBF (Gauss)	−4.4519	−4.4519	−5.2328
CSRBF 1	−4.4567	−4.4567	−5.2327
Kriging	−4.4405	−4.4405	−5.2325

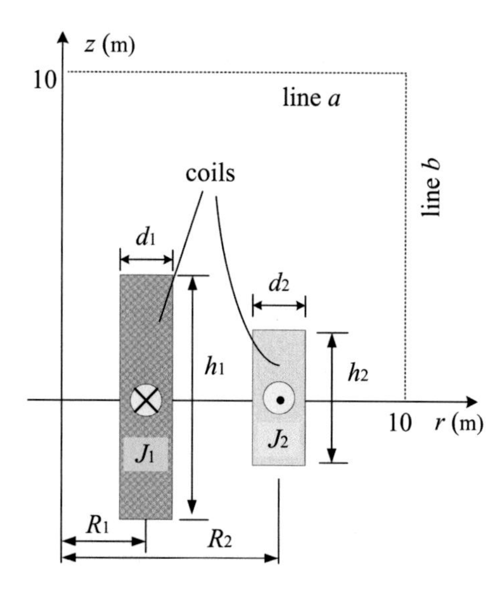

Fig. 4.3 The geometry configuration of SMES (axisymmetric)

the efficiencies of different optimization methods [12, 13, 15–19]. This problem consists of several cases, including low and high dimensional, single- and multi-objective, and discrete as well as continuous parameter optimization cases. Figure 4.3 illustrates the optimization structure of this SMES. As shown, it consists of two solenoids and there are eight design parameters.

For this benchmark problem, there are three optimization objectives as listed below:

(1) The expect value of the stored energy, E, in this SMES is 180 MJ;
(2) The mean stray fields, B_{stray}, should be as small as possible; The value of B_{stray} can be calculated by $B_{\text{stray}} = \left(\sum_{i=1}^{21} \frac{|B_{\text{stray}}^i|^2}{21}\right)^{1/2}$, where B_{stray}^i is the magnetic flux density evaluated along 21 equidistant points on lines a and b, and
(3) The magnetic field should maintain the superconducting condition of the storage. The superconducting material employed in this SMES is NbTi. Figure 4.4 illustrates the critical curve of this material. To ensure the superconducting condition, the current density, J_i, in the solenoids and magnetic field density in the storage must follow the following constraint

$$|J_i| \leq -6.4|B_{\max}|_i + 54.0 \tag{4.4}$$

In the optimization case of discrete parameters, the dimensions of the inner solenoid, R_1, $h_1/2$, and d_1, are fixed at the values as shown in Table 4.5, and the current densities, J_1 and J_2, are fixed at 22.5 A/mm^2. The dimensions of the outer solenoid, R_2, $h_2/2$, and d_2, are to be optimized to reduce the stray fields while keeping the stored energy close to 180 MJ. The optimization model is defined as

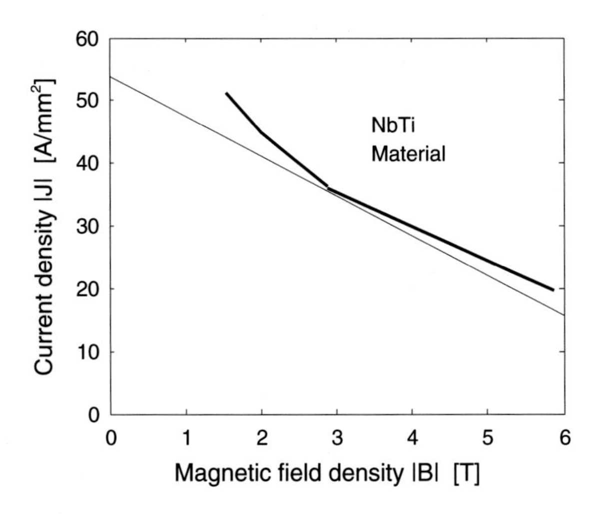

Fig. 4.4 Critical curve of superconducting material

Table 4.5 Design parameters and values of SMES for discrete case

Par.	Unit	Min	Max	Step size	fixed
R_1	m	–	–	–	2.0
R_2	m	2.6	3.4	0.01	–
$h_1/2$	m	–	–	–	0.8
$h_2/2$	m	0.204	1.1	0.007	–
d_1	m	–	–	–	0.27
d_2	m	0.1	0.4	0.003	–
J_1	A/mm^2	–	–	–	22.5
J_2	A/mm^2	–	–	–	22.5

$$\begin{aligned} &\min : f(\mathbf{x}) = B_{\text{stray}}/B_{\text{norm}} \\ &\text{s.t.} \begin{cases} h(\mathbf{x}) = |E/180 - 1| = 0 \\ g(\mathbf{x}) = B_{\max} - 4.92 \leq 0 \end{cases}, \end{aligned} \tag{4.5}$$

where B_{norm} = 3 mT, $h(\mathbf{x})$ and $g(\mathbf{x})$ are two constraints, and $g(\mathbf{x})$ is an inequality constraint concerning the quench condition that guarantees superconductivity. As the current density is fixed at 22.5 A/mm^2, the corresponding $B_{\max}$ is 4.92 T obtained by (4.4). In the optimization, the constraints are maintained by using a penalty function defined as

$$F(\mathbf{x}) = f(\mathbf{x}) + 1000[h(\mathbf{x})^2 + \max(g(\mathbf{x}), 0)^2]. \tag{4.6}$$

Table 4.6 lists the optimization results for this SMES by using different optimization methods, including direct optimization method and SOM based on different approximate models. The parameters used in SOM are $\delta = \varepsilon = 2.5\,\%$ and

Table 4.6 Optimization results of SMES

Par.	R_2	$h_2/2$	d_2	E	B_{max}	B_{stray}	F	FEM
Unit	m	m	m	MJ	T	mT	–	–
TEAM	3.08	0.239	0.394	179.86	4.73	0.9084	0.3034	–
DEA	3.18	0.428	0.211	180.00	3.83	1.0323	0.3441	2310
RSM	3.08	0.246	0.382	179.68	4.70	0.9051	0.3049	164
RSM-MLS	3.10	0.274	0.337	179.93	4.53	0.9171	0.3058	171
RBF-Gauss	3.16	0.365	0.244	179.95	3.94	0.9573	0.3192	202
CSRBF1	3.11	0.267	0.340	179.94	4.50	0.9431	0.3145	157
Kriging	3.11	0.267	0.340	179.94	4.50	0.9431	0.3145	157

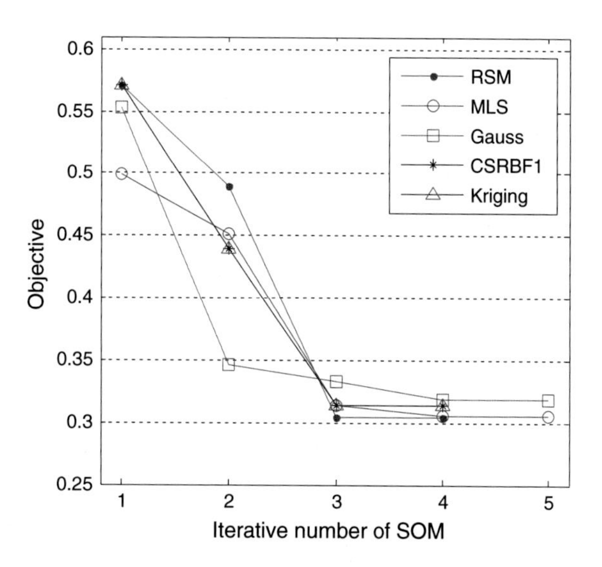

Fig. 4.5 Convergence rates of SOM by using different models

$N^{(1)} = [5, 5, 3]$. The best solution of this problem given by the TEAM workshop is also listed in the table. Figure 4.5 illustrates the convergence rates of SOM by using different approximate models. The first three iteration processes are for the COPs. The last one or two iteration processes are for the FOPs. The following conclusions can be drawn from the table and figure:

(1) By using the direct optimization method based on FEM and DEA, the obtained optimal result is 180.00 MJ for the stored energy (slightly better than the TEAM value, 179.86 MJ), 3.83 T for the maximum magnetic flux density, and 1.0323 mT for the mean stray flux density (slightly higher than the TEAM value, 0.9084 mT). The objective is 0.3441, which is much higher than the TEAM objective, 0.3034. To obtain this optimal result, 2310 FEM samples are used.
(2) By using the SOM based on RSM, the obtained optimal result is 179.68 MJ for the stored energy, 0.9051 mT for the mean stray flux density, and 0.3049 for the objective. The obtained objective is lower than that obtained by the

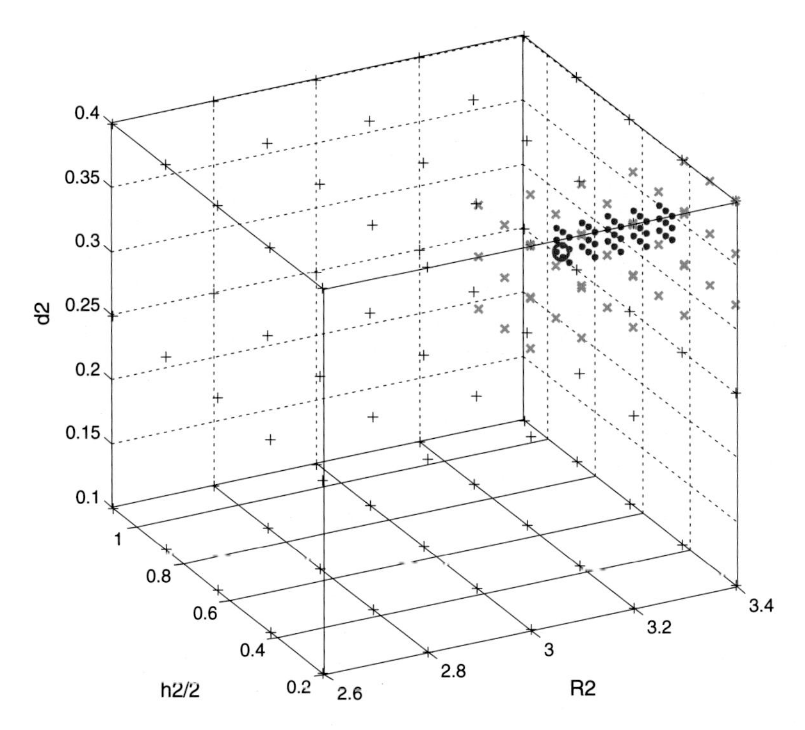

Fig. 4.6 Total FEM samples required by RSM-SOM

direct optimization method and close to the TEAM value. Most importantly, this method requires only 164 FEM samples to obtain the optimal result. Figure 4.6 depicts these samples using "+", "×" and "•" to denote the points sampled from three COPs, and "o" the points sampled in the FOP. As shown, the samples are distributed non-uniformly in the whole design space, and the sampling processes approach the final optimal solutions step by step.

Figure 4.7 illustrates the magnetic flux density distribution in the SMES for this optimal solution. As shown, the maximal flux density is 4.702 T, which is very close to 4.70 T, the value listed in Table 4.6.

Therefore, SOM based on RSM has high optimization efficiency while maintaining the accuracy of optimum.

(3) The detailed optimization results of SOM based on other models are shown in Table 4.6. As shown, all optimal objectives obtained from SOM with different models are slightly higher than the TEAM value and smaller than that of DEA. Both the SOM based on CSRBF1 model and the SOM based on Kriging model used only 157 FEM samples, which is 6.8 % (=157/2310) of that used by the DEA method.

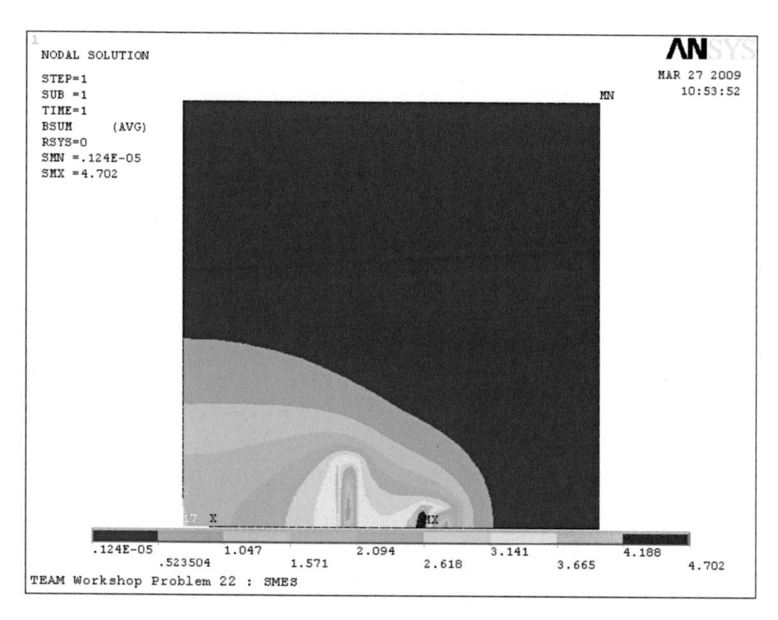

Fig. 4.7 Field analysis by ANSYS for SMES

In summary, all solutions obtained by the SOM based on various models are better than that of the direct optimization or DEA method, and there is not much difference between the approximate models. In other words, SOM has high optimization efficiency whereas the model type does not affect its optimization accuracy very much. Unlike the case presented in Chap. 3 that MLS is better than LSE for RSM and CSRBF is better than RBF, they do not have much difference within the framework of SOM. On the other hand, the MLS and CSRBF require extra parameters, which make them more complex compared with RSM, RBF and Kriging. Therefore, RBF and Kriging are two good models for SOM.

4.3.4 Improved SOM

As mentioned in the above section, space reduction strategy plays an important role in SOM. The design purpose of the former space reduction strategy is to minimize the distance between the mean of next design range and the optimal result [12]. It is accurate from the point of view of distance minimization, but it has not considered the issue that reduces the number of FEM samples effectively. To make full use of most points sampled in the last set, we present a new space reduction strategy in this section.

Assume $x^{(k)} = [x_{li}^{(k)}, x_{ui}^{(k)}]$ is the boundary of the *i*th variable in the *k*th optimization process, $l^{(k)}$ the interval, $h^{(k)}$ the step size, $N^{(k)}$ the number of sample points, $S^{(k)}$ the sample set, and $x_o^{(k)}$ and $f^{(k)}$ are the optimal result and corresponding function value, respectively. The new space reduction strategy consists of the following two steps.

Reduction step:

$$\hat{x}_{li}^{(k+1)} = \max\left\{x_{li}^{(k)}, round\left(\frac{x_{oi}^{(k)} - \Delta l}{\Delta h}\right)\Delta h\right\}, \tag{4.7}$$

$$\hat{x}_{ui}^{(k+1)} = \min\left\{x_{ui}^{(k)}, round\left(\frac{x_{oi}^{(k)} + \Delta l}{\Delta h}\right)\Delta h\right\}. \tag{4.8}$$

Correction step:

$$x_{li}^{(k+1)} = x_{li}^{(k)} + round\left(\frac{2(\hat{x}_{li}^{(k+1)} - x_{li}^{(k)})}{h_i^{(k)}}\right)\frac{h_i^{(k)}}{2}, \tag{4.9}$$

$$x_{ui}^{(k+1)} = x_{li}^{(k)} + round\left(\frac{2(\hat{x}_{ui}^{(k+1)} - x_{li}^{(k)})}{h_i^{(k)}}\right)\frac{h_i^{(k)}}{2}, \tag{4.10}$$

where *round*(x) is a function to round x to its nearest integer, $\Delta l = l_i^{(k)}/n_l$, $\Delta h = h_i^{(k)}/n_h$, and n_l and n_h are the two reduction factors with defaults 4 and 8, respectively [20].

To check the efficiency of the new method, a comparison with the former space reduction strategy is conducted.

As an example, assume that the initial design space is [0, 1], N = 6, and the uniform sampling method is used. The first sample data is then $S^{(1)}$ = {0.0, 0.2, 0.4, 0.6, 0.8, 1.0}, and the optimal value is supposed as 0.35. By the former space reduction strategy, the next design space is [0.1, 0.6], and the next sample set $S^{(2)}$ = {0.1, 0.2, 0.3, 0.4, 0.5, 0.6}. Thus, 3 sample points have been sampled in $S^{(1)}$, or 50 % computation cost is saved. On the other hand, by the new space reduction strategy, the next sample space is [0.2, 0.6], and $S^{(2)}$ = {0.2, 0.3, 0.4, 0.5, 0.6}, resulting in a save of 60 % computation cost, which is better than the former strategy.

As another example, if the optimal value is 0.3, the next sample space is [0.05, 0.55] by the former strategy, and $S^{(2)}$ = {0.05, 0.15, 0.25, 0.35, 0.45, 0.55}. Thus, no sample point has been sampled. By the new strategy, however, the next sample space is [0.0, 0.6], and $S^{(2)}$ = {0.0, 0.1, 0.2, 0.3, 0.4, 0.5, 0.6}. Thus, 4 sample points have been sampled in the $S^{(1)}$, resulting in a reduction of 57.14 % computation cost.

Therefore, the new strategy is more effective than the former one.

Table 4.7 shows mean saving rates of sample points for the former and new space reduction strategies by using MCA. For each strategy and every sample

Table 4.7 Mean saving rates by two strategies of SOM

N	3	4	5	6
Former	0.2293	0.2531	0.2428	0.2375
New	0.5415	0.5383	0.5334	0.5221

Table 4.8 Optimization results of SMES by using improved SOM

Par.	Unit	DEA	Gauss	MQ
R_2	m	3.18	3.12	3.07
$h_2/2$	m	0.428	0.309	0.295
d_2	m	0.211	0.295	0.328
E	MJ	180.00	179.94	179.64
B_{stray}	mT	1.0323	0.9300	0.9657
B_{max}	T	3.83	4.31	4.60
F	–	0.3441	0.3101	0.3259
FEM	–	2310	214	129

number N, 10^6 random numbers are generated as the current optimal points by Monte Carlo method. Then, the mean saving rate is obtained for each case. As shown, all saving rates by the new strategy are more than 50 %, which are higher than those obtained by the former strategy [20].

Table 4.8 lists the optimization results of the SMES by using two kinds of RBF models. As shown, the Gauss RBF model gives the best result, and the multi-quadrics (MQ) RBF model requires the least number of FEM samples. Furthermore, either RBF model requires less than 1/10 of the FEM samples needed by DEA.

4.3.5 A PM Claw Pole Motor with SMC Stator

A three-phase PM claw pole motor is investigated in this section to demonstrate the efficiency of the improved SOM. Figure 4.8 shows the stator part and FEM region for this motor [21, 22]. It can be seen that 3D FEM is required for the performance analysis of this motor. The stator of this motor is made of SMC material. Figure 4.9 shows a molded SMC claw pole disk. This motor was designed to deliver a power of 60 W at 3000 rev/min to replace an existing single-phase induction motor in a dishwasher pump. Table 4.9 lists the design dimensions of this motor.

Figure 4.10 illustrates the prototype of this claw pole motor fabricated with the dimensions shown in Table 4.9. Figure 4.11 shows the measured motor speed against DC link voltage with different loads. It is found that the estimated performance parameters calculated from the FEM-based method agreed well with the experimental results, for example, the inductance. More details can be found in [21]. Therefore, it is reliable to use the FEM to optimize the investigated motor.

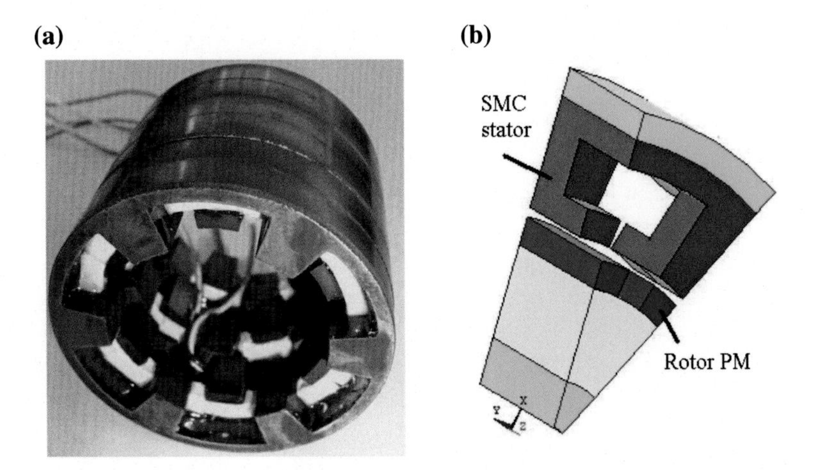

Fig. 4.8 **a** Molded stator of a PM claw-pole motor, and **b** FEM region of a pole of a stack

Fig. 4.9 Molded SMC claw-pole disk

Table 4.9 Main design dimensions and variables

Par.	Description	Unit	Value
–	Number of poles	–	12
R_{so}	Stator outer radius	mm	33.5
R_{si}	Stator inner radius	mm	21.5
b_s	Width of side wall	mm	6.3
h_{rm}	Radial length of magnet	mm	3.0
ρ	SMC core's density	g/cm^3	5.8
g_1	Air gap	mm	1.0
h_p	Claw pole height	mm	3.0
h_{sy}	Stator yoke thickness	mm	3.0
N_c	Number of winding turns	turn	256

Regarding the optimization of this motor, the objective is chosen to minimize the material cost while maximizing the output power or torque at 3000 rev/min. The outer radius and axial length, or volume, of the motor is fixed in the optimization. The material cost mainly includes the costs of PM, copper, SMC core, and steel.

Fig. 4.10 Experimental setup for the test of prototype

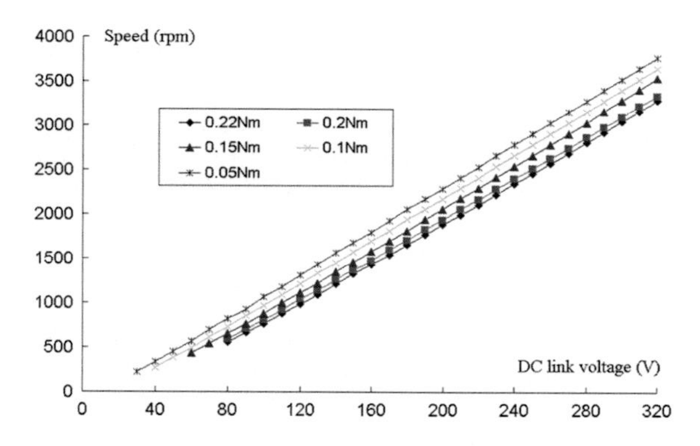

Fig. 4.11 Speed versus DC link voltage with constant torque

Three constraints are also considered. The optimization model can be defined as the following

$$
\begin{aligned}
\text{min:} \quad & f(\mathbf{x}) = \frac{Cost}{C_0} + \frac{P_0}{P_{\text{out}}} \\
\text{s.t.} \quad & g_1(\mathbf{x}) = 0.78 - \eta \leq 0 \\
& g_2(\mathbf{x}) = 60 - P_{\text{out}} \leq 0 \\
& g_3(\mathbf{x}) = J_c - 4.5 \leq 0
\end{aligned}
\tag{4.11}
$$

where C_0 and P_0 are the cost and output power of the initial prototype, 0.78 and 60 the rated values of efficiency (η) and output power (P_{out}), respectively for the initial design, and the last constraint is the current density, J_c, of winding, which should be no more than 4.5 A/mm^2 in terms of its specifications [22].

From previous design experience, three parameters R_{si}, b_s and h_{rm}, are important to the motor performance. Therefore, they are selected as the optimization variables. Table 4.10 lists the optimization results by using improved SOM and Kriging model. As shown, the obtained output power and material cost are 98 W and $8.99, which are better than those of the initial design (60 W and $14.18). Meanwhile, only 197 FEM samples are reuqired by improved SOM to obtain the optimal

Table 4.10 Optimization results of the SMC claw pole motor

Par.	Unit	Initial	SOM
R_{si}	mm	21.50	18.50
b_s	mm	6.30	4.00
h_{rm}	mm	3.00	2.00
η	%	78	83
P_{out}	W	60	98
Torque	Nm	0.19	0.31
Cost	$	14.18	8.99
f_m	–	2.00	1.25
FEM	–	–	197

solution. Compared with the samples required by direct optimization method, for example, DEA requiring about 3 × 5 × 170 or 2550 FEM samples to obtain the optimization results, where 3 × 5 is the population size and 170 the average iteration number, 92.27 % of FEM computation cost has been saved. Therefore, the improved SOM is efficient for low-dimensional electromagnetic design problems.

4.4 Multi-objective Sequential Optimization Method

Multi-objective design optimization problems of electrical machines and electromagnetic devices have attracted great amount of research interests recently. Various design examples and benchmark problems have been proposed, such as TEAM Problems 22 (SMES) and 25 (die-press model) [17, 23–25]. To deal with these problems, many evolutionary multi-objective optimization algorithms have been employed, such as MOGA and multi-objective particle swarm optimization (MPSO) algorithm.

These algorithms have been proven efficient by many test functions and engineering examples, e.g. the TEAM workshop problems. The greatest advantage of these methods is that the designer can obtain a set of Pareto optimal solutions by a single run. For a practical design requirement, one only needs to choose the best from the obtained Pareto solutions, rather than to run the algorithm again. Thus, these algorithms can improve the computational efficiency for engineering applications with the obtained Pareto solutions.

For practical design optimization of industrial electromagnetic devices, the implementation process can be usually very time-consuming because of the use of FEM which takes most of the optimization time, especially for some complex electromagnetic devices, e.g. 3D flux PM motors. Therefore, how to efficiently employ these algorithms to deal with the multi-objective design problems of industrial electromagnetic devices of complex structures is still an open problem [23, 25, 26]; and not much work has been reported in the literature.

To address this problem, an alternative method is to use approximate models. Many kinds of approximate models, which are widely used in single-objective optimization, have been investigated in multi-objective optimization problems as well, such as RBF and Kriging models [25]. For this approach, the optimization efficiency depends highly on the model accuracy, which in turn depends on the sampling method and model type. As mentioned previously, the high level and full factor sampling method has been used in many researches because of its capability of producing an accurate solution. However, its computational cost may be expensive in many situations. Meanwhile, it should be noted that there are some key differences between single- and multi-objective optimizations based on approximate models. Each objective or constraint in the multi-objective optimization has its own characteristic, such as linear or nonlinear, convex or non-convex, and maximal or minimal value, and thus how to ensure the same modeling accuracy for all models is a key issue.

A multi-objective sequential optimization method (MSOM) is presented to deal with these problems in this work. A test function and a 3D PM TFM will be investigated to show the efficiency of the proposed method.

4.4.1 Method Description

Generally, a multi-objective optimization model has the form as

$$\begin{array}{ll} \min: & \{f_1(\mathbf{x}), f_2(\mathbf{x}), \ldots, f_p(\mathbf{x})\} \\ \text{s.t.} & g_i(\mathbf{x}) \le 0,\ i = 1, 2, \ldots, m \\ & \mathbf{x}_l \le \mathbf{x} \le \mathbf{x}_u,\ \mathbf{x} = [x_1, \ldots, x_D]^T \end{array}, \tag{4.12}$$

where p, m and D are the numbers of objectives, inequality constraints and variables, respectively.

In general, the solutions of a multi-objective optimization problem can be illustrated as a Pareto optimal set, and its front is not a point, but a continuous or non-continuous curve or surface. Thus, we should pay more attention to the subspace as shown in Fig. 4.12 which includes all these Pareto points and seek for a new method to sample more points in this subspace rather than the total design

(a)

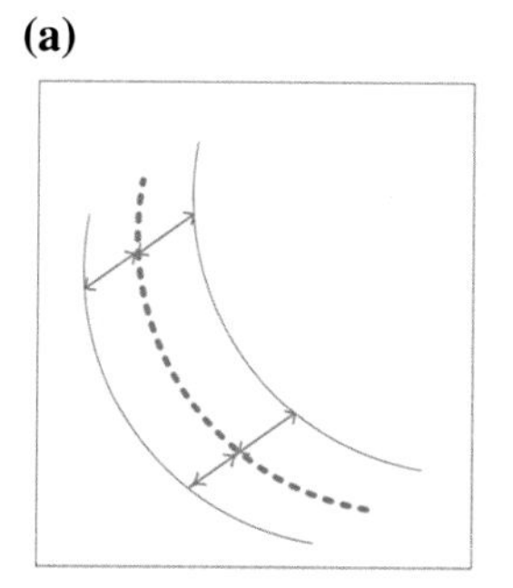

(b)

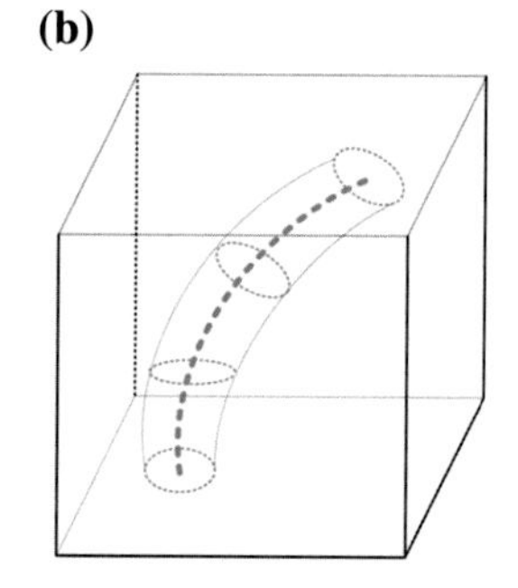

Fig. 4.12 Design idea of MSOM, **a** 2D illustration, **b** 3D illustration

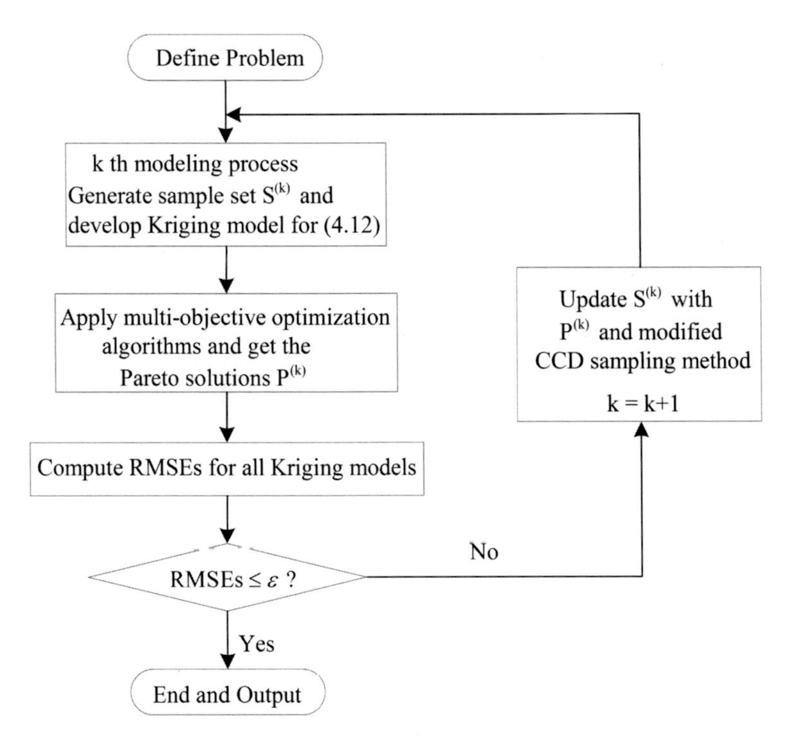

Fig. 4.13 Flowchart of MSOM

space. This approach may improve the modeling efficiency because it includes the investigation of model characteristic.

Figure 4.13 shows the flowchart of MSOM, which mainly includes the following three steps [23]:

(1) Generate an initial sample set $S^{(0)}$, and construct Kriging models for all FEMs in the design optimization problem (4.12) to get a Kriging multi-objective optimization model.
(2) Optimize the Kriging multi-objective optimization model with a multi-objective optimization algorithm, for example, NSGA II, and get the Pareto optimal solutions $P^{(k)}$. Then compute the root mean square error (RMSE) of the obtained Pareto points for each model. If all the RMSEs are no more than a constant ε, output the solutions; otherwise go to the next step.
(3) Update the sample set $S^{(k)}$ and Kriging model. As the constructed models are getting more and more accurate through the optimization process, the true Pareto solutions are probably located around the current $P^{(k)}$. To improve the modeling efficiency, a modified central composite design (CCD) sampling method is presented to update the sample sets.

The CCD is a classic sampling method for the construction of RSM. It divides the samples into two subsets, one for the property estimation of the linear term, and the

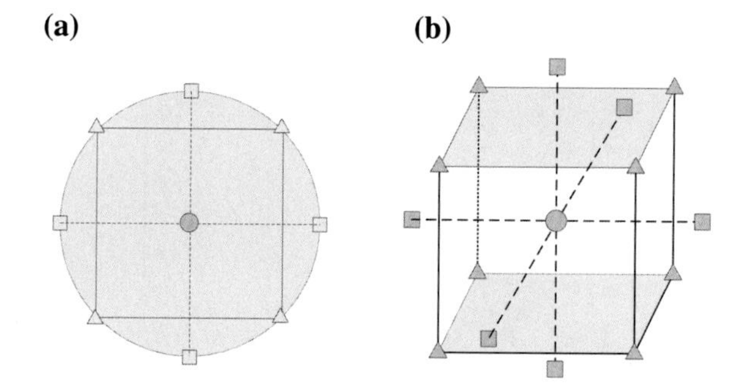

Fig. 4.14 Illustrations of the modified CCD sampling method, **a** 2D, and **b** 3D

other for the curved surface. It is claimed to be superior in the modeling of RSM [27]. Considering that RSM was used as the determined term in the Kriging model, a modified CCD sampling method has been presented in our previous work. Figure 4.14 shows two illustrations of the proposed method for (a) a two-variable case, and (b) a three-variable case. The circle points in the figure are the Pareto points. The triangle points (with number of 2^D) are sampled by the two levels full factor design method. The square points (with number of $2D$) in the axial direction are the peaks of circumscribed circle or sphere. The relationship between the diameter (d) of the circle (sphere) and the side length (l) of the square (cube) is $d = l\sqrt{D}$. For the kth optimization step, the side length is defined as half of the step size in the last step.

4.4.2 Example 1—Poloni (POL) Function

POL function investigated in Chap. 3 is a classic test function for multi-objective optimization methods as its Pareto optimal solutions are not continuous and non-convex [23, 28]. It is rewritten as (4.13). Figure 4.15 illustrates the two objectives of this function.

$$\min: \begin{cases} f_1(x_1, x_2) = 1 + (A_1 - B_1)^2 + (A_2 - B_2)^2 \\ f_2(x_1, x_2) = (x_1 + 3)^2 + (x_2 + 1)^2 \end{cases}, \tag{4.13}$$

$$\begin{aligned} A_1 &= 0.5 \sin 1 - 2 \cos 1 + \sin 2 - 1.5 \cos 2 \\ A_2 &= 1.5 \sin 1 - \cos 1 + 2 \sin 2 - 0.5 \cos 2 \\ B_1 &= 0.5 \sin x_1 - 2 \cos x_1 + \sin x_2 - 1.5 \cos x_2 \\ B_2 &= 1.5 \sin x_1 - \cos x_1 + 2 \sin x_2 - 0.5 \cos x_2 \\ &-\pi \le x_1,\ x_2 \le \pi \end{aligned}.$$

In the optimization, a controlled elitist multi-objective genetic algorithm (a variant of NSGA II) in MATLAB is used in this example. Except for the population size, all other parameters use the default values, e.g. the default Pareto fraction is 0.35, and the default value of ε in MSOM is 1 %.

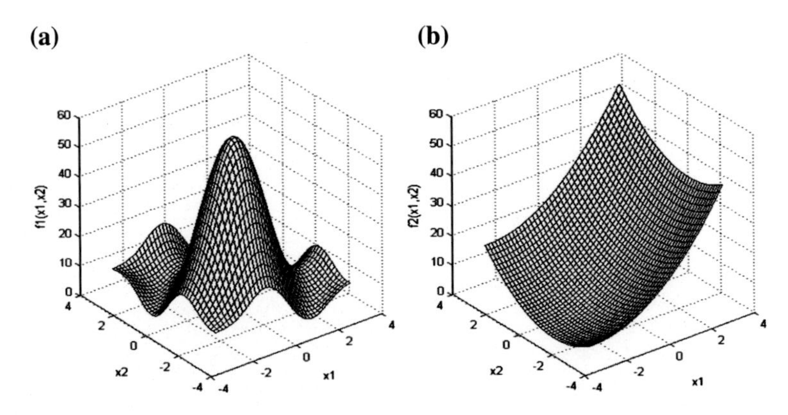

Fig. 4.15 POL function: f_1 (*left*) and f_2 (*right*)

Figure 4.16 illustrates the Pareto optimal solutions obtained by two methods, the direct function optimization with NSGA II and the proposed MSOM. For the latter case, only one model updating process is needed to get the final Pareto solutions. As shown, the Pareto solutions of this function are separated to two parts. It is non-continuous on the whole. The Pareto front of the 2nd Kriging model ($k = 2$ in MSOM) fits that from the true function very well. Figure 4.17 illustrates the total sample points sampled by MSOM which includes only 109 points. As shown, the obtained samples are non-uniformly distributed in the whole space and include more points in the subspace to which the Pareto points belong, so that the sampling efficiency can be improved by the proposed method.

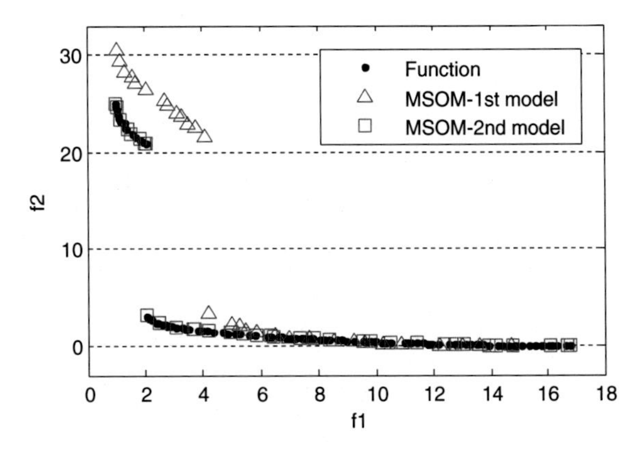

Fig. 4.16 Pareto solutions of POL function

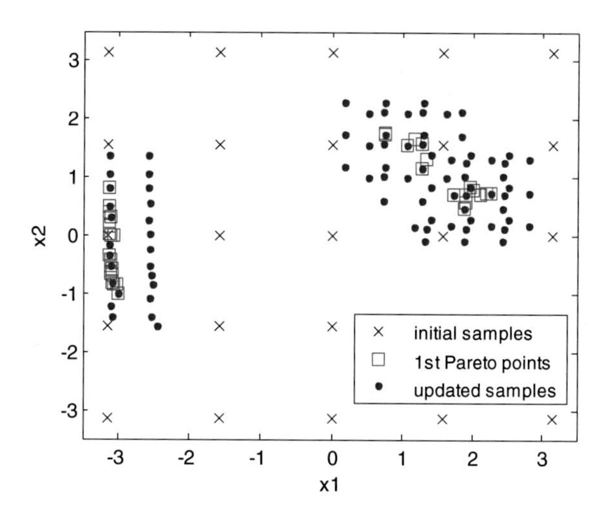

Fig. 4.17 Sample points by using MSOM

4.4.3 Example 2—A PM Transverse Flux Machine

In our previous work, a PM TFM with soft magnetic composite (SMC) stator core was developed [1, 29–32]. A multi-objective design optimization is conducted here for the broad industrial applications of this machine. A PM-SMC TFM prototype was illustrated previously in Fig. 2.3, Chap. 2 This machine was initially designed to deliver an output power of 640 W at 1800 rev/min. Table 2.1 tabulates the main dimensions. Figure 2.6 illustrates the FEM model used in ANSYS. The computation of FEM of this machine is very time-consuming as 3D FEM is required for the performance evaluation.

To get reliable analysis and optimization results, the analysis model based on FEM (as shown in Fig. 2.6) should be verified by experimental results. Figures 4.18, 4.19, 4.20 and Table 2.2 show the calculated and measured key parameters for this machine. Figure 4.18 shows the measured motor speed against output torque with different DC link voltages. Figure 4.19 illustrates the measured electromotive force (EMF) waveforms at 1800 rev/min. The measured motor back EMF constant is 0. 244 Vs, which is 1 % lower than the calculated value of 0.247 Vs. The calculated

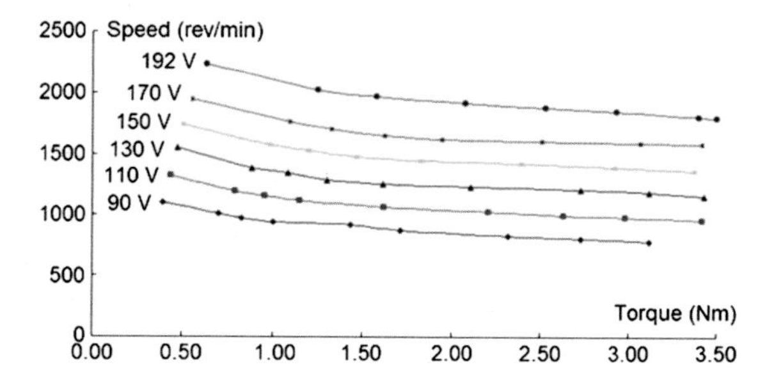

Fig. 4.18 Speed against output torque with different DC link voltages

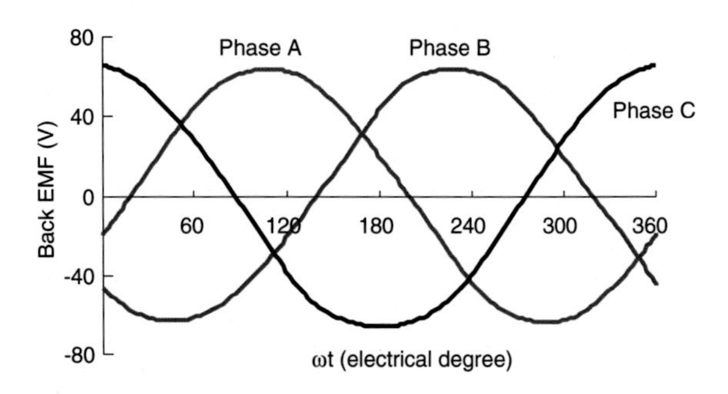

Fig. 4.19 Measured line-to-neutral back EMF

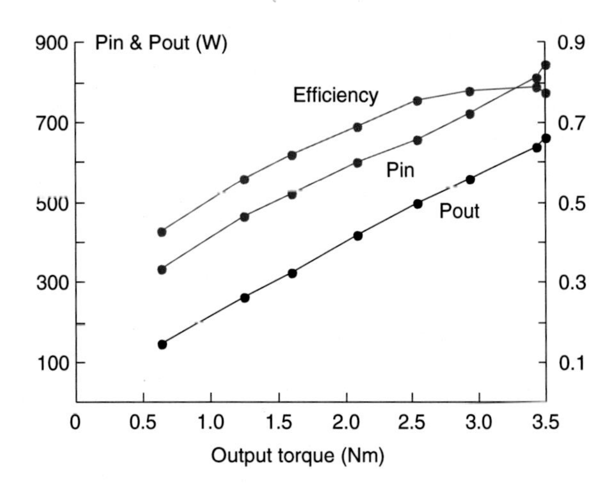

Fig. 4.20 Measured curves for input, output powers and motor efficiency in terms of torque

phase resistance and inductance, and maximal cogging torque are 0.310 Ω, 6.68 mH and 0.339 Nm, respectively, which are very close to the measured values of 0.305 Ω, 6.53 mH and 0.320 Nm. Figure 4.20 shows the measured curves of the input power, output power, and efficiency against the output torque. It is found that the estimated performance parameters calculated from FEM-based method are well aligned with the experimental results, such as the inductance and cogging torque. Therefore, all the experimental results have verified the effectiveness of this FEM-based analysis method, and it is reliable to be used for optimizing the electrical machine under investigation.

The multi-objective optimization model of this machine can be defined as

$$\begin{aligned} \min: &\quad \begin{cases} f_1(\mathbf{x}) = \text{Cost(PM)} + \text{Cost(Cu)} \\ f_2(\mathbf{x}) = 640 - P_{\text{out}} \end{cases} \\ \text{s.t.} &\quad \begin{cases} g_1(\mathbf{x}) = 0.795 - \eta \le 0, \\ g_2(\mathbf{x}) = 640 - P_{\text{out}} \le 0, \\ g_3(\mathbf{x}) = sf - 0.8 \le 0, \\ g_4(\mathbf{x}) = J_c - 6 \le 0. \end{cases} \end{aligned} \tag{4.14}$$

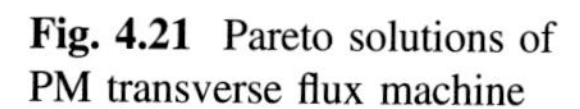

Fig. 4.21 Pareto solutions of PM transverse flux machine

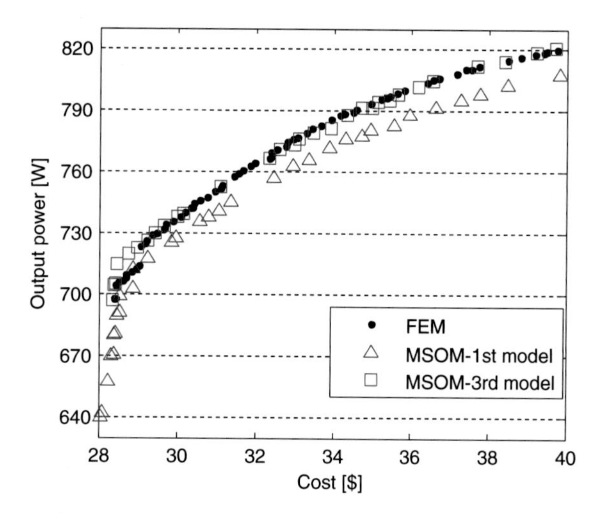

where η and P_{out} are the efficiency and output power of the machine, respectively, sf is the winding fill factor, and J_c in A/mm^2 is the current density of the copper wire. The first objective cost mainly consists of the costs of PM and copper winding. Four parameters are selected as the optimization variables in this work. They are circumferential angle and axial width of PM, and the number of turns and diameter of copper wire winding. These are the significant parameters for the objectives from our previous design experience [30].

For this problem, three model updating processes ($k = 3$ in MSOM) are needed for the MSOM to get the final Pareto solutions as shown in Fig. 4.21, which includes the initial and the last Pareto points of MSOM. Figure 4.21 also illustrates the Pareto points obtained by the direct optimization of FEM with NSGA II. As shown, the Pareto front from the MSOM fits that from FEM very well. Moreover, the needed FEM sample points of MSOM are only 556, which is about 6 % that by the direct multi-objective optimization of FEM, in which about 10,000 FEM samples are needed [23].

4.5 Sensitivity Analysis Techniques

In high dimensional design optimization problems, some design parameters relate to the objectives more significantly than others. Ignoring this fact and optimizing all design parameters in a single level (at the same time) may result in huge computing cost. For example, the optimization process of a motor with 10 parameters (dimension $D = 10$) by using the GA and FEM with the population size of 50 ($5 \times D$) and iteration number of 200 requires about 10,000 (50×200) samples, which can be a huge computational burden for many motors, especially those requiring 3D FEM.

On the other hand, it is impossible to replace the FEM with approximate models, such as RSM and Kriging model, because they cannot approximate highly dimensional problems with sufficient accuracy by using reasonably small number of samples. For example, the first step in the construction of approximation models is to use the DOE technique to obtain the initial samples. If 5 samples are required for each parameter, in total, 5^{10} FEM samples are required, which is greater than those required by direct optimization method of GA and FEM.

Therefore, the traditional direct optimization method based on FEM and the approximation models cannot solve the highly dimensional design optimization problems. To solve these problems, a multi-level optimization method was presented for electrical machines and drive systems and other electromagnetic devices in our previous work [22, 33, 34]. The main idea of the multi-level optimization method is that the high dimensional design space can be divided into two or several low dimensional design spaces in terms of the order of their sensitivities. The detailed discussion of multi-level optimization method will be presented in the next section and Chaps. 5 and 6. This section presents a brief investigation for sensitivity analysis (SA) techniques. In general, there are four types of techniques for the significance analysis of parameters in the design of PM motors. They are the sizing equation [35–37], local sensitivity analysis (LSA) [38], global sensitivity analysis [39–42], and analysis of variance (ANOVA) [22, 27] techniques, respectively. The last one, ANOVA, is based on the DOE technique. Two of them, LSA and DOE will be introduced in the following sections.

4.5.1 Local Sensitivity Analysis

Assume that $f(\mathbf{x})$ is the objective function (such as output power, torque and cost) to be optimized. Mathematically, the sensitivity of the ith parameter, x_i, at the point $\mathbf{x_0}$ can be defined as

$$S_i = \left.\frac{\partial f(\mathbf{x})}{\partial x_i}\right|_{\mathbf{x}=\mathbf{x}_0} \tag{4.15}$$

where S_i is the sensitivity. The larger the $|S_i|$, the more sensitive the objective function $f(\mathbf{x})$ is to the parameter x_i [22, 38].

It should be noted that an analytical expression of objective function is required in (4.15). However, there is no analytic form of objective function if motor's performance is calculated by using FEM. In this case, a differential form of (4.15) should be used to calculate the sensitivity as the following

$$S_i = \frac{f(\mathbf{x}_0 \pm \Delta\mathbf{x}_i) - f(\mathbf{x}_0)}{\pm\Delta x_i} \tag{4.16}$$

where Δx_i is the increment of parameter x_i. In general, there are two methods to determine this increment. The first one is known as the parameter variation method, in which Δx_i is usually defined as 10, 20 % or both of its initial value. The other method is known as the deviation variation method, in which Δx_i is usually defined as the standard deviation of x_i . In this work, as the parameter's deviation is not given, the first method will be used to calculate the sensitivity.

As an example, let us consider the output power as the objective and the dimensions of PMs, such as width, x_1, and height, x_2, as the design parameters. The sensitivity of PM width can be calculated as the following. Firstly, apply ±10 % perturbations to the PM width, and calculate the objective function (the output power) corresponding to the two samples, i.e. $(1 - 10\ \%)x_1$ and $(1 + 10\ \%)x_1$. Secondly, calculate the relative errors of these two samples by comparing the objective function of these two samples to that obtained from the initial reference point ($\mathbf{x}_0$). Finally, the average of absolute values of these two relative errors is taken as the sensitivity of PM width on the output power.

It should be noted that different parameters have different units. To make the obtained sensitivity values comparable, a normalization step below is needed

$$SS_i = \frac{\partial f(\mathbf{x})/f(\mathbf{x})}{\partial x_i/x_i} = \frac{\partial f(\mathbf{x})}{f(\mathbf{x})}\frac{1}{\delta} \approx \frac{\Delta f(\mathbf{x})}{f(\mathbf{x})}\frac{1}{\delta} \tag{4.17}$$

where δ is the ratio of changed amplitude of parameter $x_{i.}$ By taking this normalization, only the ratios of $\frac{\Delta f(\mathbf{x})}{f(\mathbf{x})}$ are compared to acquire the sensitivity value for each of the design parameters.

4.5.2 Analysis of Variance Based on DOE

Basically, DOE is a kind of statistical method which has been widely used in the design and data analysis of experiments in many areas, such as experiments in agriculture, chemistry, and industrial design. The main aim of DOE is to arrange an efficient experiment with smaller number of experiments, shorter experimental cycle, and lower experimental cost, so as to obtain good experimental results and scientific analysis conclusions. There are two types of DOE techniques, the full factor design and partial factor design. The latter includes many further types, such as orthogonal design and Latin hypercube design [22, 27].

ANOVA is a technique based on DOE, which can be used to determine the significant factors from all the design parameters. To implement the ANOVA, an experiment table should be designed firstly by using DOE. The full-factor design and orthogonal design are the two most popular DOE techniques. Because of the high dimensional feature of motor design, it is time consuming to use the full-factor method. For example, a motor design optimization problem of 8 parameters would need 2^8 or 256 samples if a two-level full-factor design scheme is used. However,

Table 4.11 The orthogonal design table of $L_9(3^4)$

No of exp.	Par. 1	Par. 2	Par. 3	Par.4
1	1	1	1	1
2	1	2	2	2
3	1	3	3	3
4	2	1	2	3
5	2	2	3	1
6	2	3	1	2
7	3	1	3	2
8	3	2	1	3
9	3	3	2	1

only 12 samples are needed if the orthogonal design is used. Similarly, 3^8 or 6,561 samples are needed if a three-level full-factor design scheme is used, whereas if we use the orthogonal design, only 27 points are needed. Therefore, orthogonal design should be a good choice for most motor design optimization problems.

To implement the orthogonal design, the first step is to select an orthogonal design table from available tables. Table 4.11 shows an orthogonal design table of $L_9(3^4)$, where subscript 9 indicates the number of experiments, and 3 the levels for each parameter, while superscript 4 indicates that this table can be used for a problem with no more than 4 design parameters and no interactions between them. Then, the numbers 1, 2, and 3 in the table are the corresponding levels for each parameter. Table 4.12 illustrates an orthogonal design table of $L_{12}(2^{11})$, which can be used for a problem with no more than 11 design parameters and no interactions between them. Similarly, there are many available orthogonal design tables, such as $L_{12}(2^{41})$, $L_{16}(4^5)$, and $L_{27}(3^{13})$.

Table 4.12 The orthogonal design table of $L_{12}(2^{11})$

No of exp.	Parameter no.										
	1	2	3	4	5	6	7	8	9	10	11
1	1	1	1	1	1	1	1	1	1	1	1
2	1	1	1	1	1	2	2	2	2	2	2
3	1	1	2	2	2	1	1	1	2	2	2
4	1	2	1	2	2	1	2	2	1	1	2
5	1	2	2	1	2	2	1	2	1	2	1
6	1	2	2	2	1	2	2	1	2	1	1
7	2	1	2	2	1	1	2	2	1	2	1
8	2	1	2	1	2	2	2	1	1	1	2
9	2	1	1	2	2	2	1	2	2	1	1
10	2	2	2	1	1	1	1	2	2	1	2
11	2	2	1	2	1	2	1	1	1	2	2
12	2	2	1	1	2	1	2	1	2	2	1

After the DOE step, ANOVA can be used to determine the significance of each parameter. ANOVA is a family of multivariate statistical technique for helping infer whether there are real differences between the means of three or more groups or variables in a population, based on the sample data. In order to determine whether the differences are significant, ANOVA is concerned with differences between the samples, known as the variance. By comparing the variance among sample members, the differences are considered to be significant if the variance is larger between samples. Therefore, ANOVA can be regarded as a statistical test that looks for significant differences between means.

The understanding of ANOVA requires the background of multivariate statistics. Fortunately, its implementation is very simple and can be realized by various software packages, such as SPSS, Minitab, Matlab, and Excel.

4.5.3 Example Study—A PM Claw Pole Motor

This section presents an example for the sensitivity analysis of a PM claw pole motor investigated in SOM. Six design parameters listed in Table 4.9 (R_{si}, b_s, h_{rm}, g_1, h_p and h_{sy}) will be investigated for the sensitivity analysis of this motor by using LSA. Table 4.13 lists the samples needed for the data analysis of LSA. For each parameter, its initial value and four variation amplitudes, −20, −10, 10, and 20 % will be considered. Totally, 25 samples are needed for the calculation, which includes 24 points for those four variations (−20, −10, 10, and 20 %) of six parameters and 1 initial sample ("0" column in the table).

Table 4.14 tabulates the analysis data obtained from the LSA technique. For the sake of comparison, an average column, i.e. mean sensitivity, is listed in this table. The sensitivity order can be obtained from the data in this column as

$$|b_s| > |h_{rm}| > |R_{si}| > |g_1| > |h_p| > |h_{sy}| \tag{4.18}$$

To balance the optimization framework, we can take three of them as the significant factors, which are R_{si}, b_s, and h_{rm}. Actually, they are the parameters used for the analysis of improved SOM in Sect. 4.3.5 [22].

Table 4.13 Samples for LSA

Par.	Amplitude variations of parameter (δ)				
	−20 %	−10 %	0	10 %	20 %
R_{si}	17.2	19.35	21.5	23.65	25.8
b_s	5.04	5.67	6.30	6.93	7.56
h_{rm}	2.40	2.70	3.00	3.30	3.60
g_1	0.80	0.90	1.00	1.10	1.20
h_p	2.40	2.70	3.00	3.30	3.60
h_{sy}	2.40	2.70	3.00	3.30	3.60

Table 4.14 Sensitivity analysis data for claw pole motor

Par.	Amplitude variations of parameter (δ)					Sensitivity
	−20 %	−10 %	0	10 %	20 %	
R_{si}	0.002	−0.012	0	0.027	0.066	0.0267
b_s	−0.071	−0.045	0	0.079	0.244	0.1095
h_{rm}	−0.100	−0.051	0	0.051	0.101	0.0754
g_1	−0.018	−0.010	0	0.009	0.020	0.0141
h_p	0.011	0.004	0	−0.003	−0.004	0.0053
h_{sy}	−0.003	−0.002	0	0.001	0.002	0.0019

4.6 Multi-level Optimization Method

4.6.1 Method Introduction

Figure 4.22 illustrates the flowchart of the multi-level optimization method or sequential subspace optimization method. This method is mainly proposed to improve the optimization efficiency of high dimensional design optimization problems of electrical machines and other electromagnetic devices. It consists of the following main steps [1, 22, 33, 34].

Step 1: Define the optimization problems, including objectives, constraints, design parameters and their ranges.

Step 2: Implement sensitivity analysis for all design parameters, and obtain the order of sensitivities of them.

Step 3: Divide the initial high dimensional design space into two or three low dimensional subspaces in terms of the sensitivity order of design parameters obtained in Step 2. Consider the situation of three subspaces as an example. The first subspace (X1) includes all highly significant factors, and the second subspace (X2) consists of all significant factors, while the third subspace (X3) all non-significant factors.

Step 4: Optimize the highly significant factor subspace X1. In the implementation, the initial parameters in X2 and X3 are fixed.

Step 5: Optimize the significant factor subspace X2. In the implementation, the parameters in X1 are fixed at the solutions from Step 2, and the parameters in X3 are fixed at those in Step 4.

Step 6: Optimize the non-significant factor subspace X3. In the implementation, the parameters in X1 and X2 are fixed at the solutions from the last two steps.

Step 7: If the objective meets the specification, output the optimal solutions. Otherwise, update the parameters in X2 and X3, and go to Step 4 and conduct the optimization again till convergence.

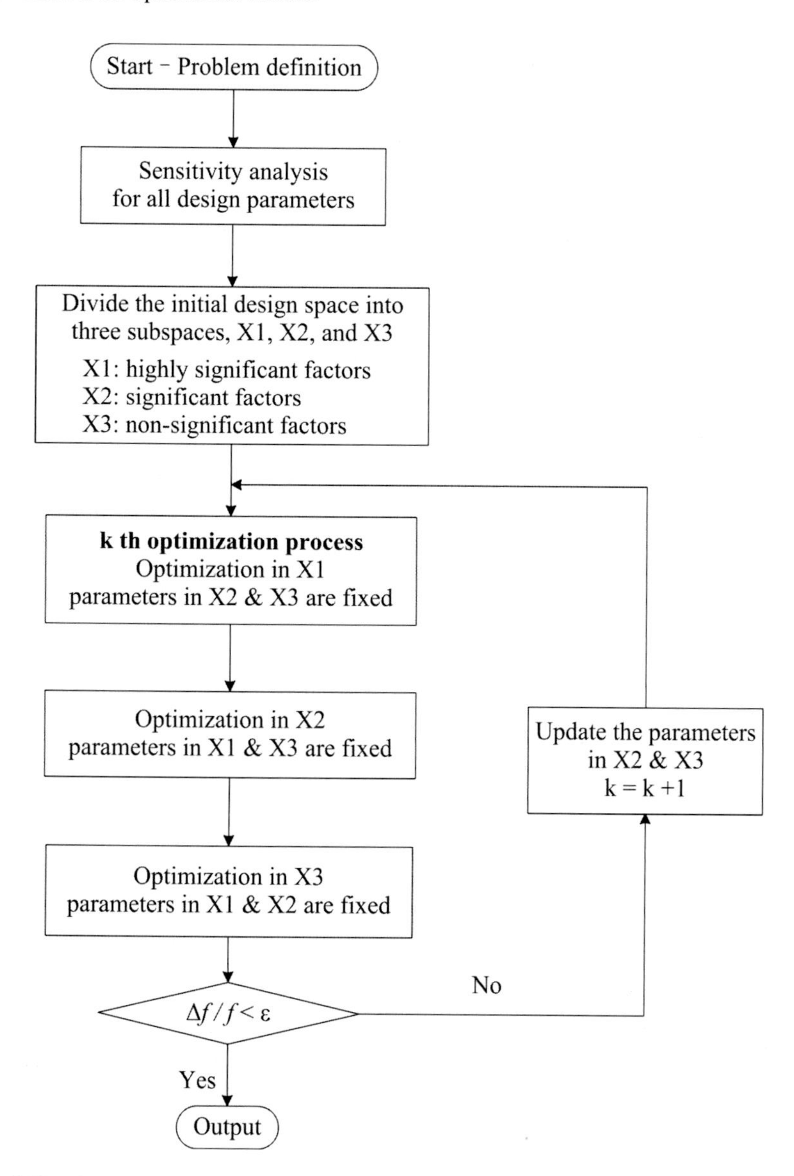

Fig. 4.22 Flowchart of multi-level optimization method

It should be noted that two subspaces maybe reasonable for some problems. In that case, combine X1 and X2 into one subspace, which may be known as the significant factor subspace. Meanwhile, the dimension of subspace is much smaller than that in the initial space. Thus, the traditional direct optimization methods and approximate models can be used in each of them.

4.6.2 Example Study—SMES

Figure 4.3 illustrated a benchmark design example about the SMES. A three parameter discrete optimization case has been investigated in Sect. 4.3.3. This section will investigate an eight parameter continuous case. Table 4.15 lists the scope of these eight parameters. From the previous discussion, it can be found that those eight parameters can be divided into two subspaces in terms of their significances. One subspace involves the four parameters of inner coil $\{R_1, h_1, d_1, J_1\}$, and the other one includes the other four parameters of outer coil $\{R_2, h_2, d_2, J_2\}$ [33].

Table 4.16 shows obtained optimization solutions for this continuous case of SMES. As shown, 4720 FEM samples are needed for DEA to get the optimal solution under the direct optimization framework. The optimal stored energy in SMES is 178.75 MJ (the error is 1.25 MJ), and mean stray field is 2.27 mT.

By the multi-level optimization method, only 1078 FEM samples are needed for the optimization, which is less than 1/4 of that by DEA. The resultant optimal mean stray field is 3.23 mT, slightly higher than that by DEA, and the error of energy is 1.01 MJ which is smaller than that given by DEA. Therefore, the proposed

Table 4.15 Design parameters of SMES under continuous case

Parameter	Unit	Min.	Max.
R_1	m	1.0	4.0
R_2	m	1.8	5.0
$h_1/2$	m	0.1	1.8
$h_2/2$	m	0.1	1.8
d_1	m	0.1	0.8
d_2	m	0.1	0.8
J_1	A/mm^2	10.0	30.0
J_2	A/mm^2	10.0	30.0

Table 4.16 Optimization results for SMES under continuous case

Var.	Unit	DEA	Multi-level
R_1	m	2.382	2.662
R_2	m	3.377	4.015
$h_1/2$	m	1.118	1.049
$h_2/2$	m	0.366	0.421
d_1	m	0.188	0.223
d_2	m	0.653	0.368
J_1	A/mm^2	22.57	18.09
J_2	A/mm^2	11.06	11.08
B_{stray}	mT	2.27	3.23
E	MJ	178.75	178.99
FEM	–	4720	1078

multi-level optimization method is more efficient than DEA for high dimensional design problems of electromagnetic devices. It will be employed to optimize electrical machines and drive systems in the following two chapters.

4.7 Multi-level Genetic Algorithm

4.7.1 Problem Matrix

The aforementioned multi-level optimization method is based on the space division strategy. The following multi-level optimization method is mainly based on a kind of intelligent optimization algorithm called multi-level genetic algorithm (MLGA). It presents an alternative and efficient way to implement the multi-level optimization for electrical machines as well as other electromagnetic devices and systems [43–46].

In MLGA, the relationship between the design variables, constraints, and objective functions can be described by the problem matrix, as shown in Fig. 4.23. The design variables may be assigned into different sub-vectors according to the relationships between design variables. The variables which have close relationship should be allocated to the same sub-vector.

In the figure, the symbols $P_{ij}(i = 0, 1, \ldots, n, j = 0, 1, \ldots, m)$ are the coefficients indicating the relative importance between the design variables and objective functions, as well as constraints in the correlation analysis [44]. The P value tests whether there is sufficient evidence that the correlation coefficient is not zero. The greater the P value is, the less the relative importance of the design variable to the objective function is. The samples of variables are determined by the DOE. Some commercial statistic software packages, such as SPSS and Minitab, can provide the modules for relative importance analysis.

According to the P values in the problem matrix, the design variables may be arranged on diverse levels. For one objective function, the variables possessing similar P values will be managed on the same level.

Fig. 4.23 Structure of problem matrix for MLGA

Design parameters	x_1	x_2	x_3	x_4	…	x_m
Objective function	P_{01}	P_{02}	P_{03}	P_{04}	…	P_{0m}
Constraint 1	P_{11}	P_{12}	P_{13}	P_{14}	…	P_{1m}
Constraint 2	P_{21}	P_{22}	P_{23}	P_{24}	…	P_{2m}
⋮	⋮	⋮	⋮	⋮	⋮	⋮
Constraint n	P_{n1}	P_{n2}	P_{n3}	P_{n4}	…	P_{nm}

4.7.2 Description of MLGA

The traditional GA creates a vector (chromosome) encoded by all the design variables and then applies evolution operation to all the individuals described as chromosomes in one population. In MLGA the design optimization variables are classified and allocated to different levels according to the relative importance between the variables and objective functions, constraints, as well as the practical engineering weighting factors and optimization sequence. The variables on different levels are encoded independently. Each level may have multiple populations and each of them can adopt different genetic operators and parameters. The relationship between sub-problems in multi-level problems can be handled by MLGA.

The architecture of MLGA is illustrated in Fig. 4.24. As shown, the upper level (GA_1) is the master GA module. The second (GA_{2i}) and third (GA_{3i}) consist of a number of modules (or subsystems). The GA in one subsystem will be affected by other modules. The module in the upper level of the MLGA acts not only as a solver of the corresponding sub-problem, but also as a coordinator and controller of the modules on the lower level. This means that the lower level module GA_{ij} will be affected by the upper level module $GA_{i-1,j}$, and even by the adjacent modules $GA_{i,j-1}$ and $GA_{i,j+1}$ on the same level.

The GA can be described as follows:

$$GA = (\mathbf{IP}, \mathbf{PS}, \mathbf{EL}, \mathbf{FIT}, \mathbf{SO}, \mathbf{CO}, \mathbf{MO}) \tag{4.19}$$

where **IP**, **PS**, **EL** and **FIT** represent the initial population, population size, encoding length and fitness value, respectively, **SO**, **CO** and **MO** are the genetic operations, namely, selection, crossover and mutation operations.

The MLGA can be described as follows.

$$GA_{ij} = (\mathbf{IP}_{ij}, \mathbf{PS}_{ij}, \mathbf{EL}_{ij}, \mathbf{FIT}_{ij}, \mathbf{SO}_{ij}, \mathbf{CO}_{ij}, \mathbf{MO}_{ij}) \tag{4.20}$$

where GA_{ij} stands for applying the independent GA to the ith level and the jth module. With the reaction between different levels and adjacent sub-modules on the same level, GA_{ij} can be described as the following:

$$\begin{aligned} GA_{ij} = \big(&\mathbf{IP}_{ij}(GA_{i,j-1}, GA_{i-1,j}, GA_{i,j+1}) \\ &\mathbf{PS}_{ij}(GA_{i,j-1}, GA_{i-1,j}, GA_{i,j+1}) \\ &\mathbf{EL}_{ij}(GA_{i,j-1}, GA_{i-1,j}, GA_{i,j+1}) \\ &\mathbf{FIT}_{ij}(GA_{i,j-1}, GA_{i-1,j}, GA_{i,j+1}) \\ &\mathbf{SO}_{ij}(GA_{i,j-1}, GA_{i-1,j}, GA_{i,j+1}) \\ &\mathbf{CO}_{ij}(GA_{i,j-1}, GA_{i-1,j}, GA_{i,j+1}) \\ &\mathbf{MO}_{ij}(GA_{i,j-1}, GA_{i-1,j}, GA_{i,j+1})\big) \end{aligned} \tag{4.21}$$

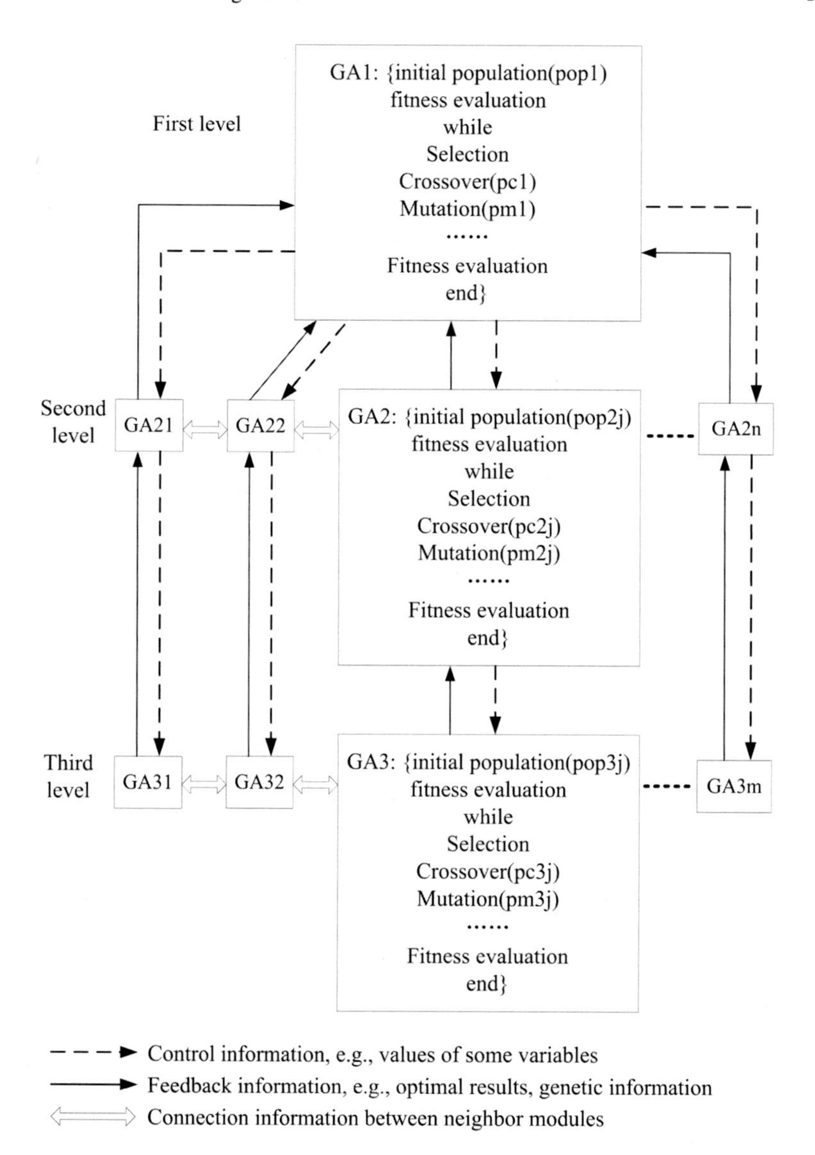

Fig. 4.24 Framework of MLGA

The MLGA can be implemented by the following steps:

Step 1: Determine the objective functions, constraints and design variables.

Step 2: Analyze the relationship of design variables, objective functions and constraints by using the correlation analysis, and construct the problem matrix.

Step 3: Determine the architecture of MLGA, including the number of levels and the number of modules in each level.

Step 4: Allocate the design variables, objective functions and constraints on different levels according to the problem matrix, and build up the relationships among different levels and different modules on each level. Each module corresponds to a genetic algorithm module.

Step 5: Implement the MLGA modules of each level from the top to the lowest level. The upper level module sends control messages and parameter values to the lower level module. Feedback messages from the lower level are used as the evaluation function by the upper level.

Step 6: The total solving process ends when the termination criterion of the top level has been reached. Otherwise, Step 5 will be repeated.

4.7.3 Example Study—SPMSM

4.7.3.1 Optimization Model of SPMSM

In this section, a surface-mounted permanent magnet synchronous machine (SPMSM) will be optimized by using the MLGA. The motor is rated with an output power of 950 W (or rated torque of 4.5 Nm) at speed 2000 rev/min and supplied by the rated line-to-line voltage of 128 V. Figure 4.25 shows two photos of the commercially manufactured motor and its name plate. Figure 4.26 shows the model for the 2D finite element analysis of this machine, and Table 4.17 lists the main structural parameters.

The stator and rotor cores are not permitted to be modified due to manufacture limitation. The coil pitch, number of parallel branches, and number of wires per conductor of the 3-phase windings are fixed. The magnet thickness *hm* and width *bm*, the diameter of conductor *WindD* and the conductors per slot *Ns* are chosen as the design optimization variables. The optimization objective is to achieve the maximum efficiency within reasonable cost of conductors and magnets. The constraints are the fill factor and rated output power. The optimization model can be described as

$$\begin{aligned} \max \quad & f(\mathbf{x}) = \frac{K}{w_1 \frac{100-\eta}{100} + w_2 \frac{Cost}{Cost_{\max}}} \\ \text{s.t.} \quad & P_2 > 945W \\ & sf < 78\,\% \end{aligned} \tag{4.22}$$

where the design variable $\mathbf{x} = [hm, bm, Ns, WindD]$, Ns and $WindD$ are discrete variables, η is the efficiency of the SPMSM, P_2 the output power, sf the fill factor, and K, w_1 and w_2 are the weighting factors specified by the designer [44].

(a)

(b)

Fig. 4.25 Photos of the SPMSM, a whole motor, b name plate

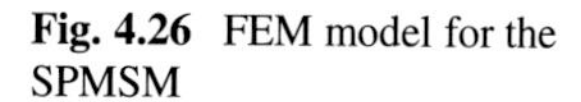

Fig. 4.26 FEM model for the SPMSM

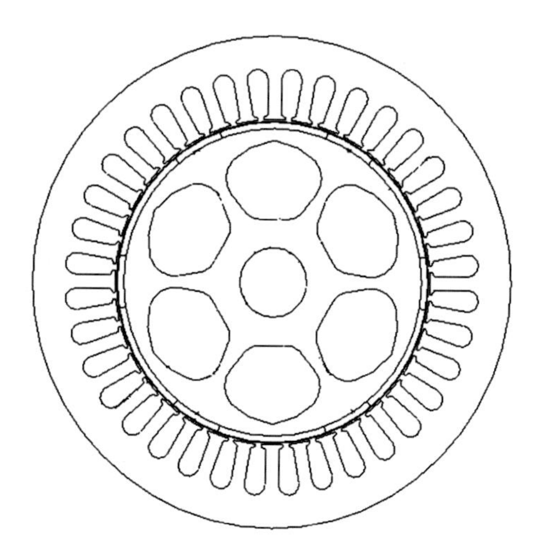

Table 4.17 Main dimensions of the SPMSM

Parameters		Unit	Values
Stator	Number of slots	–	36
	Length of stack	mm	40.83
	Length of slot	mm	41.4
	Conductors per slot	–	72
	Diameter of conductor	mm	0.5
Rotor	Number of poles	–	6
	Length of PM	mm	40.0
	Width of PM	mm	31.4
	Thickness of PM	mm	1.8
	Number of PMs per pole	–	5
	Shaft diameter	mm	19.0
Air gap	Length of air gap	mm	1

4.7.3.2 Optimization Results and Discussion

A. Determination of multi-level optimization framework

The bi-level optimization model is chosen, with the objective function and constraints (4.22) shared by both levels. The fitness functions of both levels are the same, and the penalty function method is applied to deal with the constraints.

Figure 4.27 shows the problem matrix. According to the theory of correlation analysis and DOE, the *P* values which represent the relative importance between design variables and the objective functions as well as the constraints are analyzed by Minitab, a commercial statistic software package.

As shown, the *P* values of *Ns* and *WindD* are less than those of *hm* and *bm* with respect to the objective function. Therefore, *Ns* and *WindD* are significant to efficiency and costs. *hm* and *bm* are regarded as the variables of level 1 and *Ns* and *WindD* are assigned on level 2.

B. Experimental results and FEM for no-load EMF, L_{ad} and L_{aq}

On level 1, to account for the nonlinear characteristics of the core, the quasi-static FEM is applied to calculate the no-load EMF per turn and the *d*- and *q*- axis components of per turn inductances, i.e. L_{ad} and L_{aq}, to acquire highly accurate

Parameters	*hm*	*bm*	*Ns*	*WindD*
Max $f(\mathbf{x})$	0.270	0.666	0.001	0.000
$P_2 > 945$ W	0.005	0.25	0.32	0.005
$sf < 78\%$	1.000	1.000	0.000	0.000

Fig. 4.27 Problem matrix of MLGA for SPMSM

Table 4.18 Experimental results versus FEM results

Method	Unit	Experiment	FEM
Back-EMF	V	74.0	81.0
L_{ad}	H	0.015	0.0126
L_{aq}	H	0.015	0.0124

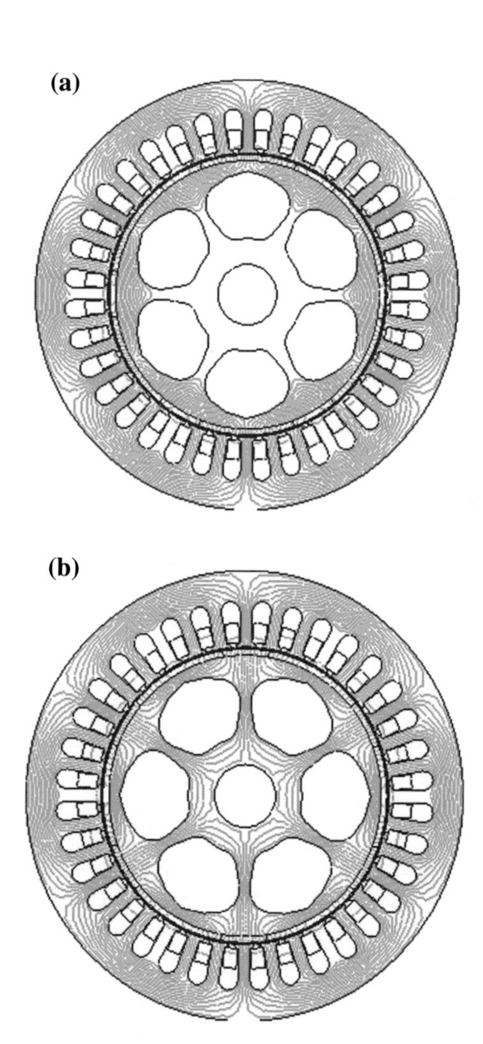

Fig. 4.28 Magnetic field distribution for calculation of **a** L_{ad} and **b** L_{aq}

parameters when the magnet thickness and width are changed. Table 4.18 lists the main motor parameters obtained from the experimental results, and the calculated results based on FEM. Figure 4.28 illustrates the magnetic field distribution when L_{ad} and L_{aq} are calculated. Figure 4.29 shows the bi-level architecture of optimization for SPMSM [44].

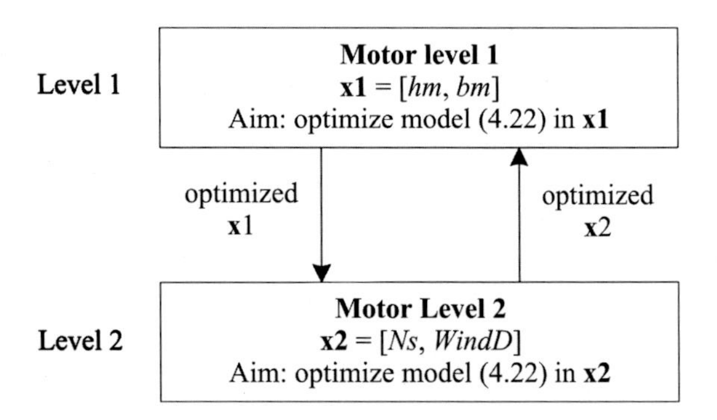

Fig. 4.29 Optimization flowchart for SPMSM

Table 4.19 Optimization results by MLGA and GA

Par.	Unit	Original	MLGA	GA
hm	mm	1.8	2.3	2.1
bm	mm	31.4	30.3	30.3
Ns	–	72	67	66
WindD	mm	0.5	0.56	0.56
η	%	83.7	86.4	86.1
Cost	$	26.1	22.6	21.5
P_2	W	946	950	951
sf	%	67	78	77

C. Comparison between MLGA and traditional GA

Both the MLGA and the traditional GA (single-level) are conducted for solving the optimization problem of SPMSM. The numbers of populations on levels 1 and 2 are 15 and 25, respectively. The number of evolution generations is 20 in each level. 40 populations and 40 evolution generations are defined in the single level GA. Table 4.19 lists the original design, the optimal results by MLGA and traditional GA.

As shown, both the MLGA and the traditional GA may achieve higher efficiencies than the original design. The efficiency optimized by the MLGA is higher than that optimized by the traditional GA. The higher the efficiency is, the higher the costs of conductors and permanent magnets will be.

Figure 4.30 illustrates the traces of fitness functions of the MLGA and the traditional GA. It can be seen that the MLGA possesses better optimal fitness values than the single-level GA. It is suggested that MLGA can provide the better design solution because the number of populations in each level may be adjusted easily. In this case study, the GA operators have the same configuration in both MLGA and single level GA. However, the designer may define appropriate GA parameters in different levels to find the satisfactory optimum [44].

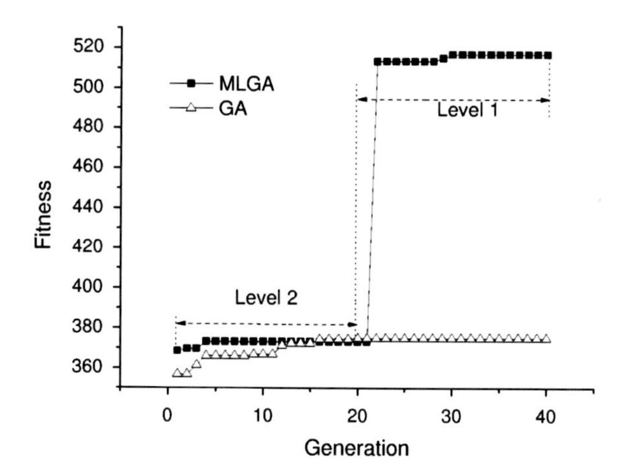

Fig. 4.30 Traces of fitness functions of MLGA and GA

4.8 Multi-disciplinary Optimization Method

4.8.1 Framework of General Multi-disciplinary Optimization

Figure 4.31 shows a classic design framework and the coupled relations for electrical machines. As shown, the design optimization is really a multi-domain problem which includes electromagnetic, material, mechanical, and thermal aspects, and they can be strongly coupled. In order to achieve high performance, the multi-disciplinary design optimization (MDO) of electrical machines must be investigated. It includes the following five main steps [47]:

Step 1: Definition of the motor specifications. It mainly consists of cost, such as material cost and manufacturing cost, output performance, such as power, torque, efficiency and speed, and other constraints, such as volume, weight, temperature rise, mechanical strength and resonance frequency, etc.

Step 2: Selection of motor type and its topology. For example, for the motor types of the PM-SMC motor investigated above, there are several options, such as TFM, claw pole and flux switching motors. The topology options may include outer rotor, inner rotor and numbers of poles for SMC motors.

Step 3: Initial design. To acquire a possible design scheme, three main designs/selections in terms of dimensions, materials and manufacturing methods are required to investigate in this step. For example, for the materials of PM-SMC motors, PM, steel, SMC and ferrite can be the options. For the manufacturing method, the press method (the moulding method) is recommended based on our design and prototyping experiences as it is widely used for batch production of PM-SMC motors.

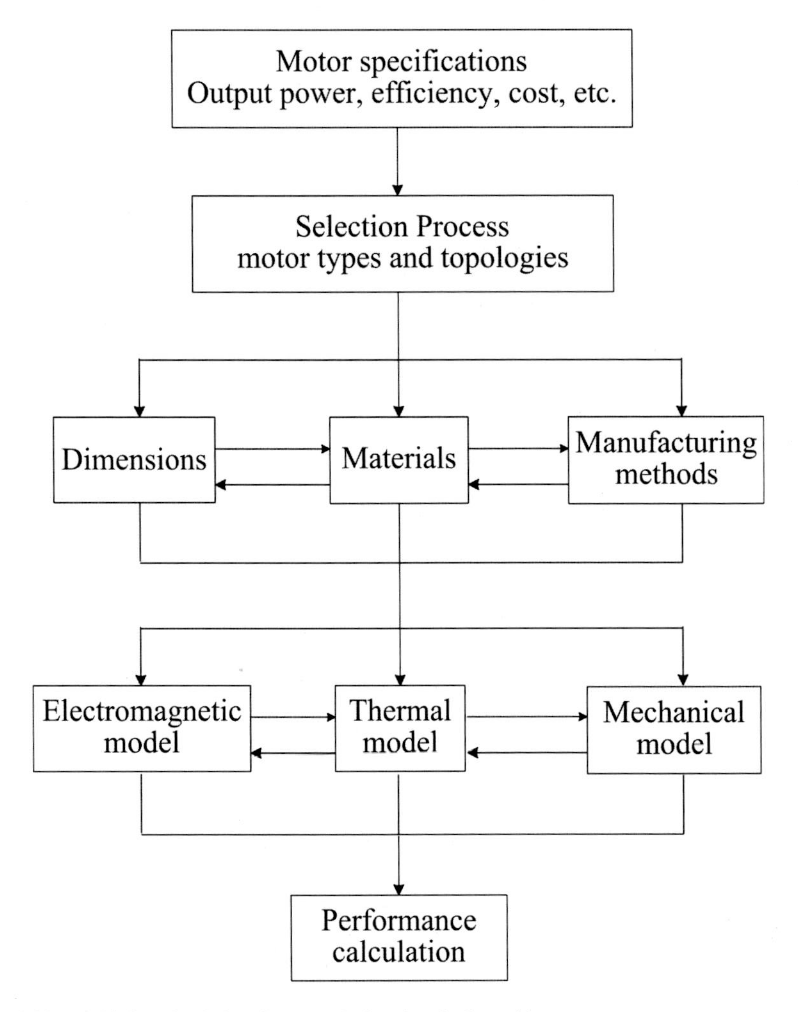

Fig. 4.31 Multi-domain design framework for electrical machines

Step 4: Development of multi-disciplinary analysis models for motors. As shown, three models are generally needed to evaluate the motor performance parameters. They are electromagnetic model including the core loss model, thermal model and mechanical model.

Step 5: Performance calculation. If the performances of the designed motor are satisfactory in terms of design specifications, this motor can be taken as an initial design scheme and can be used in the later part of optimization.

This section will take the multi-disciplinary analysis and design optimization of a PM-SMC TFM as an example to illustrate the proposed method. First of all, the electromagnetic, thermal and modal analyses are investigated for this machine based on the moulding method of the SMC cores. As modal analysis for this machine has been investigated in Sect. 2.4, the electromagnetic and thermal analyses will be the main contents in this section. Meanwhile, a lumped 3D thermal network model is developed for the thermal analysis. Then, a multi-disciplinary optimization model is proposed to minimize the material cost and maximize the output power based on the proposed thermal network model. Finally, the FEM is employed to verify the performances obtained from optimal results in terms of thermal analysis and modal analysis.

4.8.2 *Electromagnetic Analysis Based on Molded SMC Core*

As mentioned in Chap. 2, the electromagnetic analysis is mainly used to calculate the characteristic parameters of the machine, such as PM flux, core loss and inductance, so as to evaluate the performance parameters, such as output power and efficiency. On the other hand, there are two main issues which are directly related to the manufacturing of the SMC cores and will affect the electromagnetic analysis and material cost.

The first one is the mass density of SMC core. The magnetic characteristics of SMC cores depend highly on its mass density. Figure 4.32 illustrates the magnetization curves for four different densities respectively for a low density SMC core. As shown, there are significant differences between these curves and this will affect the electromagnetic analysis results.

The second one is the manufacturing cost of SMC cores. As the SMC core is compressed by a mould, the core mass density is calculated from the compacting pressure applied on the core surface and the pressure is related to press size in tons. For a given press size and dimensions of SMC core, the mass density of SMC core can be determined, and this density is directly related to the B-H curves of that core. Generally speaking, a press of larger size can produce SMC cores with higher mass density and better magnetic characteristic, but its cost is higher too. Therefore, the

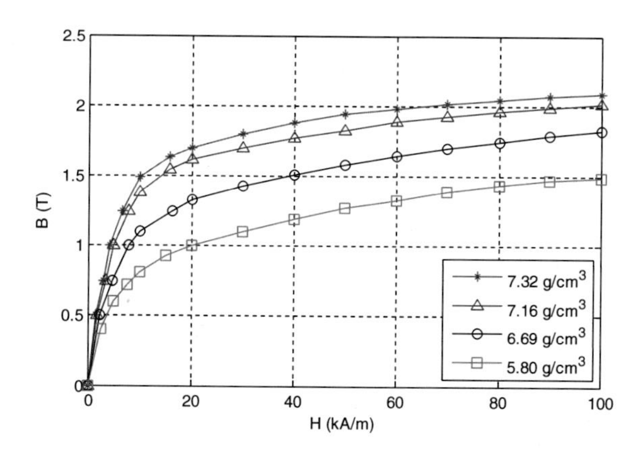

Fig. 4.32 B-H curves for three SMC density values

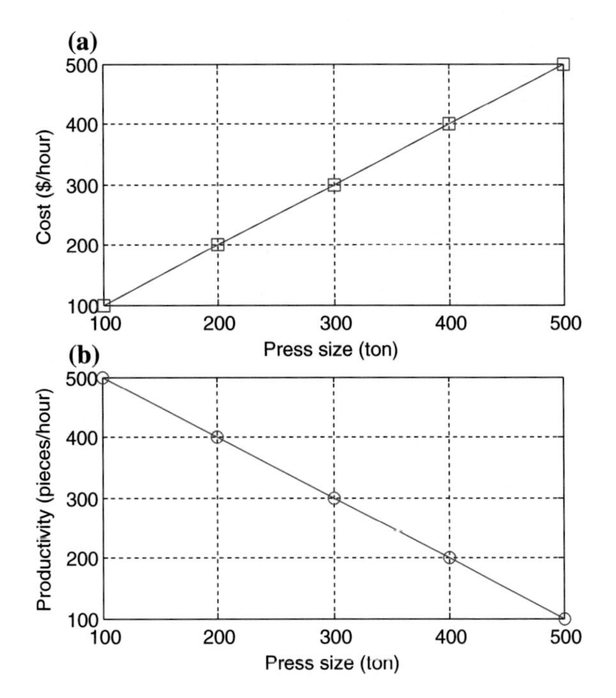

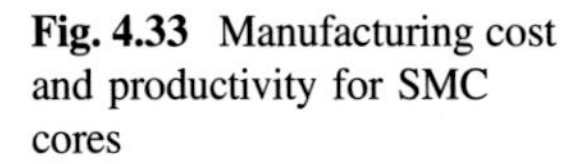
Fig. 4.33 Manufacturing cost and productivity for SMC cores

manufacturing condition is a very important factor for the design optimization of PM-SMC motors.

Figure 4.33 shows the manufacturing cost and productivity of SMC cores by using different sizes of stamping press. It can be seen that the cost is directly proportional to the press size while the productivity is inversely proportional to the press size. For example, a 100 ton press can produce 500 SMC cores per hour with a cost of $100 per hour, i.e. only $0.20 each core, while a 500 ton press can only produce 100 cores per hour with a cost of $500 per hour, meaning $5 each core. This is a big difference in industrial mass production.

4.8.3 Thermal Analysis with Lumped 3D Thermal Network Model

Thermal analysis is used to calculate the temperature rises in winding and PM rotor for this machine, so as to ensure that the motor works safely [47, 48]. For this machine, a 2D thermal network model as shown in Fig. 2.8 was developed in our previous work to simulate the thermal analysis. Considering the thermal isotropy of SMC material, we developed a 3D thermal network of lumped parameters as it can provide more accurate results than the 2D network model.

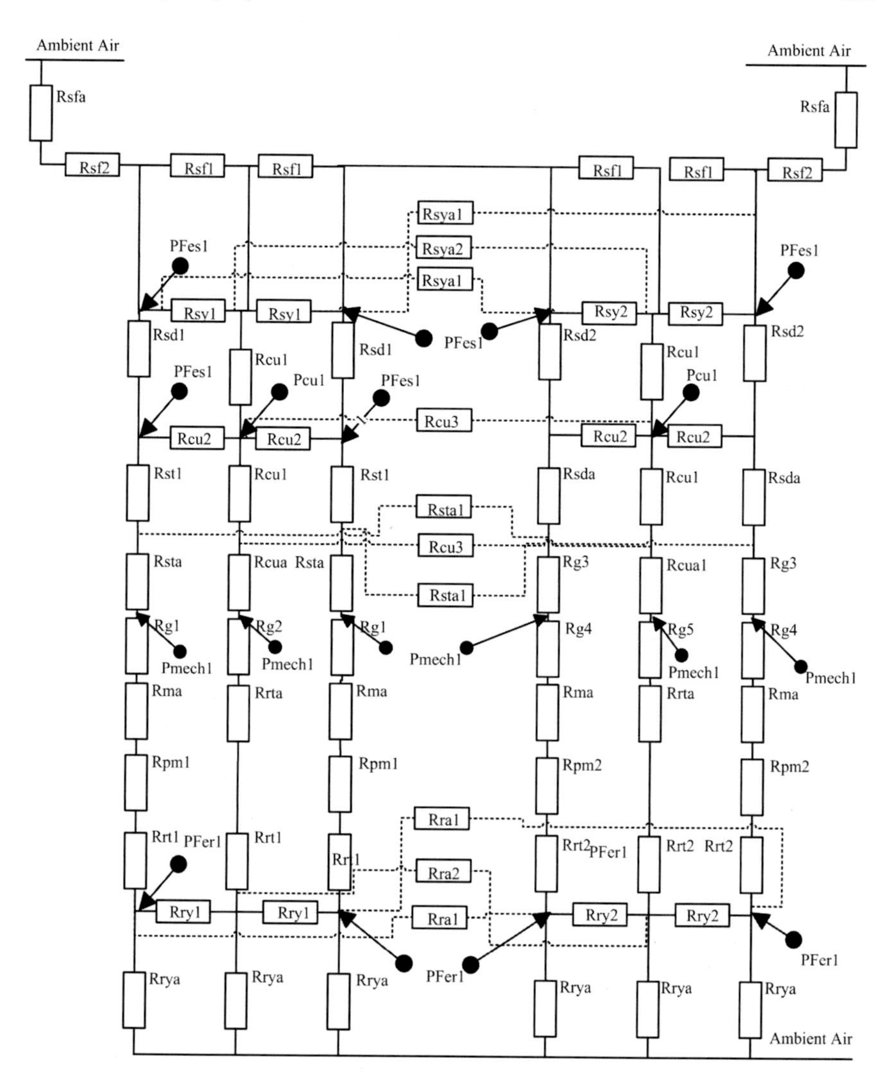

Fig. 4.34 3D thermal network model of the PM-SMC TFM

Figure 4.34 describes the 3D thermal network model for the middle stack of the studied PM TFM. The two side stacks can be neglected as the temperature of the middle stack is higher than those of two side stacks. Therefore, only the middle stack is investigated in this model. Meanwhile, the major heat dissipates from the rotor. The resistances to the thermal conduction of the following sections are calculated: two segments of stator yoke ($R_{sy1}$and R_{sy2}), stator side disk (R_{sd1} and R_{sd2}), coils (R_{cu1}, R_{cu2}, and R_{cu3}), stator teeth (R_{st1}), air gaps (R_{g1}, R_{g2}, R_{g3}, R_{g4} and R_{g5}), PMs (R_{pm1}, R_{pm2}), rotor in radial direction (R_{rt1}, R_{rt2}), rotor in axial direction (R_{ry1}, R_{ry2}), and shaft (R_{sf1} and R_{sf2}). The equivalent resistances are calculated for thermal convection between the air and stator teeth (R_{sta}), air and stator disk (R_{sda}), air and coil (R_{cua}), air and PM (R_{pm1}, R_{pm2}), air and rotor (R_{rta}), rotor and outer ambient air (R_{rya}), and shaft and air (R_{sfa}), respectively. The thermal resistances in the circumferential direction in this motor are calculated for thermal conduction in the stator yoke (R_{sya1} and R_{sya2}), coil (R_{cu3}), stator teeth (R_{sta1}), and rotor (R_{ra1} and R_{ra2}).

The heat sources in this model include the stator and rotor core losses, copper loss, and mechanical loss. In order to gain a relatively high accuracy, each loss is divided into several parts. The stator core loss is divided into six parts (P_{Fes1}), the copper loss two parts (P_{cu1}), the rotor core loss four parts (P_{Fer1}), and mechanical loss six parts (P_{mech1}), respectively.

Based on this 3D thermal network model, it is found that there is 68 °C temperature rise in the coil and 27 °C in the rotor yoke surface for this PM-SMC TFM prototype. Comparison with experimental results will be shown in Sect. 4.8.5.

4.8.4 Multi-disciplinary Design Optimization

Based on the above analysis methods, a multi-disciplinary optimization model can be developed for this PM-SMC TFM in the form as the following

$$\begin{aligned} \text{min:} \quad & f(\mathbf{x}) = \frac{Cost}{C_0} + \frac{P_0}{P_{out}} \\ \text{s.t.} \quad & g_1(\mathbf{x}) = 0.795 - \eta \le 0, \\ & g_2(\mathbf{x}) = 640 - P_{out} \le 0, \\ & g_3(\mathbf{x}) = sf - 0.7 \le 0, \\ & g_4(\mathbf{x}) = T_{PM} - 65 \le 0, \\ & g_5(\mathbf{x}) = T_{Coil} - 65 \le 0, \end{aligned} \tag{4.23}$$

where $\mathbf{x}$ is a vector of design parameters, C_0 and P_0 are the cost and output power of the initial design scheme, η and P_{out} in g_1 and g_2 the motor's efficiency and output power, respectively, sf in g_3 is the fill factor, and T_{PM} and T_{Coil} in g_4 and g_5 are the temperature rises in the PM and windings, respectively. From our design experience, six parameters as shown in Table 4.20 are significant to the performance of this machine. The cost in the objective function mainly includes the material and manufacturing costs of the SMC core.

Modal analysis is not included in this optimization model. However, to ensure that the optimized motor has good mechanical performance, modal analysis will be presented in the next section to verify the performance of the optimized motor. For the thermal analysis in optimization, the lump 3D thermal network model is used to replace the FEM analysis to improve the optimization efficiency. Then, a FEM method will be presented for the thermal analysis of the final optimal scheme in the next section.

Table 4.20 Optimization results of PM-SMC TFM

Par.	Description	Unit	Initial	MDO
–	PM circumferential angle	deg	12	10.02
–	PM width	mm	9	7.53
–	Number of turns of winding	Turn	125	110
–	Diameter of copper wire	mm	1.25	1.3
–	Air gap length	mm	1.0	0.9
ρ	Core density	g/cm^3	7.32	6.39
η	Efficiency	–	79.5 %	84.3 %
P_{out}	Output power	W	640	677
T_{PM}	Temperature rise in PM	°C	36.1	23.9
T_{Coil}	Temperature rise in coil	°C	64.9	65.0
–	Cost	$	35.8	26.5

4.8.5 *Optimization Results and Discussion*

A. Experimental verification of electromagnetic and thermal analyses
To get reliable optimization results, all analysis models should be verified by experimental results first. Figures 4.18, 4.19, 4.20 and Table 2.2 compare the calculated and measured key motor parameters for this machine. As investigated in Sect. 2.2.3, the calculated motor electromagnetic parameters are very close to the measured values.

Regarding the thermal analysis, the experimental results has shown that the temperature rises in the coil and rotor yoke are 66 and 27 °C, respectively. Compared with the calculated values (68 and 27 °C) obtained from 3D thermal network model, the maximal relative error is only 3 %. The calculated temperature rise in the coil by using the FEM method is about 63 °C, resulting in a relative error of 4.8 % compared with the experimentally measured results.

In summary, it is found that the performance parameters calculated based on both FEM and 3D thermal network model agree well with the experimental results. Therefore, these models are reliable for the optimization.

B. Optimization results
Table 4.20 lists the optimization results. By comparing these results, the following conclusions can be drawn:

(1) By the MDO, the obtained optimal cost is $26.5, the output power 677 W, and the efficiency 84.3 %, respectively. The obtained output power and efficiency are higher than those of the initial design scheme, namely 640 W and 79.5 %.
(2) The optimal SMC core density obtained by MDO is 6.39 g/cm^3, which is much smaller than that of the initial one, namely 7.32 g/cm^3. Therefore, lower

manufacturing condition and cost may be requested compared with initial design.

(3) By the thermal optimization, the temperature rises in the coil and PM from MDO are 65 and 23.9 °C, respectively.

C. FEM verification for electromagnetic, thermal and modal analyses

As only FEM model was employed for electromagnetic analysis in the optimization, FEM verifications are presented in this part for all electromagnetic, thermal and modal analyses for the obtained optimal design scheme. Figure 4.35 illustrates the electromagnetic analysis for this motor under the MDO optimum. Figure 4.36 illustrates the thermal analysis for this motor under the MDO optimum. As shown, the average temperature rise in the coil is around 62.3 °C lower (or 4.6 % relative error) than that obtained from the thermal network model.

Figure 4.37 illustrates the first-order modal analysis for this motor with MDO optimum. As shown, the resonance frequency of the optimal motor is about 4,262 Hz, which is much larger than the electromagnetic frequency of 300 Hz.

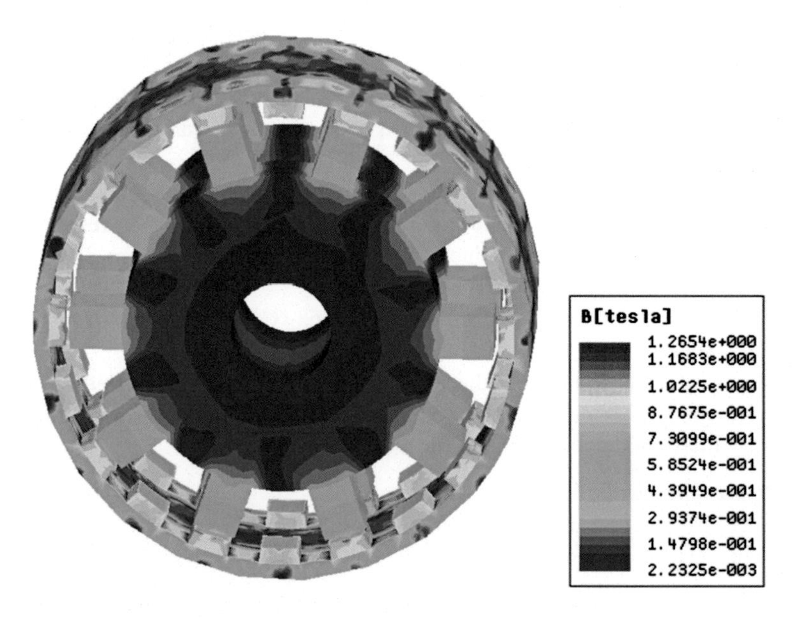

Fig. 4.35 Filed distribution for the MDO optimum

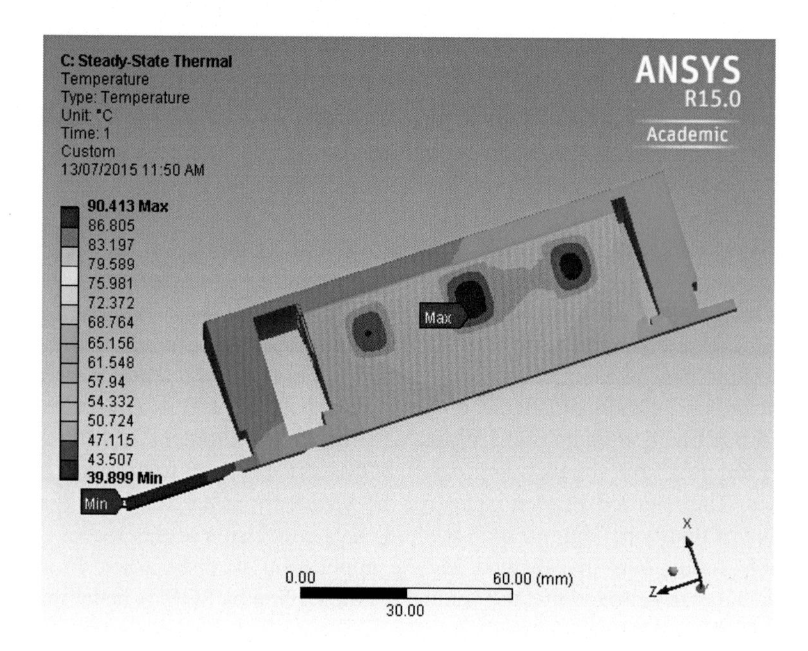

Fig. 4.36 Distribution of temperature rise in the coil for MDO optimum

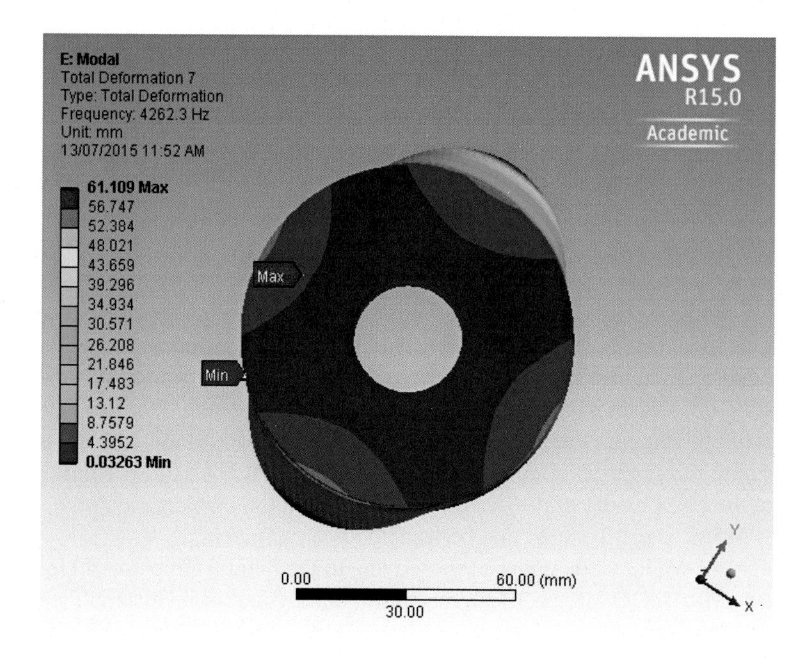

Fig. 4.37 Illustration of first order modal analysis for MDO optimum

4.9 Summary

For different design optimization problems of electrical machines with various numbers of objectives, dimensions of design parameters, and numbers of disciplines involved, different kinds of optimization methods have been presented in this chapter.

Firstly, SOM was presented to improve the optimization efficiency of low-dimensional design problems of electrical machines. Compared to the traditional direct optimization methods, SOM can be regarded as a "space-to-space" optimization method, which uses the space reduction technique to reduce the initial large design space to a small subspace around the optimal point step by step. To illustrate the optimization efficiency of the proposed method and its improvement, a standard test function and TEAM Workshop Benchmark Problem 22 (SMES) are investigated first. As shown, (improved) SOMs are efficient for these standard examples. The required FEM samples for the SMES by using SOM are less than 10 % that of traditional optimization method. As a conclusion of this section, a PM claw pole motor was investigated for the application in dishwasher. From the discussion, it can be found that the optimal solution given by SOM is better than the traditional direct optimization method in terms of output power, efficiency, material cost and optimization efficiency. Therefore, SOM is efficient for the design of low-dimensional electrical machines.

Secondly, MSOM was presented for the situation of multi-objective design optimization of electrical machines. A mathematical test function was investigated to verify the efficiency of the proposed method. Then, a design example of PM-SMC TFM was investigated to maximize the output power and minimize the material cost. As shown, the obtained Pareto front was well aligned with the one obtained by the MOGA. Most importantly, the required FEM samples are less than 10 % of that of MOGA based on FEM.

Thirdly, two kinds of multi-level optimization methods are presented for high-dimensional design optimization problems of electrical machines. The first one is based on sequential subspace optimization method. It uses the results of sensitivity analysis to divide the whole high dimensional design spaces into several low-dimensional design sub-spaces. Then, optimize these subspaces sequentially to get the final optimal results. To construct the optimization flowchart of this method, two kinds of sensitivity analysis techniques are discussed. Another kind of multi-level optimization method is based on the MLGA. It uses the problem matrix to determine the multi-level optimization framework. MLGA is also employed to optimize the FEM model of the design problem. From the investigation of a SMES and a SPMSM, it can be seen that both methods are efficient.

Finally, the MDO method was proposed due to the natural structure of electrical machines. The MDO of electrical machines mainly includes material, electromagnetic, thermal and mechanical analyses. For new kinds of materials and topologies, new manufacturing method should be investigated as well, which will affect the material performance of the material and manufacturing cost of the components of electrical machines. A design example about the PM-SMC TFM was investigated to illustrate the efficiency of the proposed optimization method. As shown, the MDO method can provide good optimal solutions which can satisfy the multi-disciplinary constraints, which are very important to the safe operation of the machine, including temperature rise and resonant frequency.

In summary, the proposed new optimization methods are efficient for the design optimization of electrical machines. It should be noted that the optimization models for the electrical machines are verified by comparing the FEM calculation results with experimental results. SOMs for single and multi-objective situations have been verified by test functions and TEAM Workshop Benchmark Problem 22 as well. Therefore, the efficiency of the proposed methods has been validated and can be employed for extensive engineering applications.

References

1. Lei G, Wang TS, Guo YG, Zhu JG, Wang SH (2014) System level design optimization method for electrical drive system: deterministic approach. IEEE Trans Ind Electron 61 (12):6591–6602
2. Vese I, Marignetti F, Radulescu MM (2010) Multiphysics approach to numerical modeling of a permanent-magnet tubular linear motor. IEEE Trans Ind Electron 57(1):320–326
3. Kreuawan S, Gillon F, Brochet P (2008) Optimal design of permanent magnet motor using multi-disciplinary design optimisation, In: proceedings of 18th international conference on electrical machines, Vilamoura. pp 1–6, 6–9 Sep 2008
4. Yao D, Ionel DM (2013) A review of recent developments in electrical machine design optimization methods with a permanent-magnet synchronous motor benchmark study. IEEE Trans Ind Appl 49(3):1268–1275
5. Di Barba P (2010) Multi-objective shape design in electricity and magnetism, Lecture Notes Elect Eng, vol 47
6. Reyes-Sierra M, Coello CAC (2006) Multi-objective particle swarm optimizers: A survey of the state-of-the-art. Int J Computat Intell Res 2(3):287–308
7. Yamazaki K, Ishigami H (2010) Rotor-shape optimization of interior-permanent-magnet motors to reduce harmonic iron losses. IEEE Trans Ind Electron 57(1):61–69
8. Barcaro M, Bianchi N, Magnussen F (2012) Permanent-magnet optimization in permanent-magnet-assisted synchronous reluctance motor for a wide constant-power speed range. IEEE Trans Ind Electron 59(6):2495–2502
9. Komeza K, Dems M (2012) Finite-element and analytical calculations of no-load core losses in energy-saving induction motors. IEEE Trans Ind Electron 59(7):2934–2946
10. Zhao W L, Lipo T A, Kwon B-I (2015) Optimal design of a novel asymmetrical rotor structure to obtain torque and efficiency improvement in surface inset PM motors, IEEE Trans Magn, 51 (3), Article no 8100704
11. Kim H-W, Kim K-T, Jo Y-S, Hur J (2013) Optimization methods of torque density for developing the neodymium free SPOKE-type BLDC motor. IEEE Trans Magn 49(5):2173–2176
12. Lei G, Shao KR, Guo Y, Zhu J, Lavers JD (2008) Sequential optimization method for the design of electromagnetic device. IEEE Trans Magn 44(11):3217–3220
13. Lei G, Zhu JG, Guo YG, Zou Y (2014) State of art of sequential optimization strategies for the design of electromagnetic devices. In Proceedings of the 17th international conference on electrical machines and systems (ICEMS), pp 706–709, Oct. 22–25, 2014

14. Storn R, Price K (1997) Differential evolution - a simple and efficient heuristic for global optimization over continuous spaces. J Global Optim 11:341–359
15. Takahashi RHC, Vasconcelos JA, Ramírez JA, Krahenbuhl L (2003) A multi-objective methodology for evaluating genetic operators. IEEE Trans Magn 39(3):1321–1324
16. Campelo F, Guimaraes FG, Igarashi H, Ramirez JA (2005) A clonal selection algorithm for optimization in electromagnetics. IEEE Trans Magn 41(5):1736–1739
17. Alotto P, Baumgartner U, Freschi F, Köstinger A, Magele Ch, Renhart W, Repetto M (2008) SMES optimization benchmark extended: introducing Pareto optimal solutions into TEAM22. IEEE Trans Magn 44(6):1066–1069
18. Wang LD, Lowther DA (2006) Selection of approximation models for electromagnetic device optimization. IEEE Trans Magn 42(2):1227–1230
19. Lei G, Shao KR, Guo Y, Zhu J, Lavers JD (2009) Improved SOM for high dimensional electromagnetic optimization problems. IEEE Trans Magn 45(10):3993–3996
20. Lei G, Yang GY, Shao KR, Guo YG, Zhu JG, Lavers JD (2010) Electromagnetic device design based on RBF models and two new sequential optimization strategies. IEEE Trans Magn 46(8):3181–3184
21. Guo YG, Zhu JG, Dorrell D (2009) Design and analysis of a claw pole PM motor with molded SMC core. IEEE Trans Magn 45(10):582–4585
22. Lei G, Liu CC, Zhu JG, Guo YG (2015) Techniques for multi-level design optimization of permanent magnet motors. IEEE Trans Energy Conver 30(4):1574–1584
23. Lei G, Shao KR, Guo YG, Zhu JG (2012) Multi-objective sequential optimization method for the design of industrial electromagnetic devices. IEEE Trans Magn 48(11):4538–4541
24. Lebensztajn L, Coulomb JL (2004) TEAM workshop problem 25: A multi-objective analysis. IEEE Trans Magn 40(2):1402–1405
25. Xie DX, Sun XW, Bai BD, Yang SY (2008) Multi-objective optimization based on response surface model and its application to engineering shape design. IEEE Trans Magn 44(6):1006–1009
26. Di Barba P (2009) Evolutionary multi-objective optimization methods for the shape design of industrial electromagnetic devices. IEEE Trans Magn 45(3):1436–1441
27. Wu CFJ, Hamada MS (2000) Experiments: Planning, Analysis Parameter Design Optimization. Wiley, New York
28. Deb K, Pratap A, Agarwal S, Meyarivan T (2002) A fast and elitist multi-objective genetic algorithm: NSGA-II. IEEE Trans Evol Comput 6(2):182–197
29. Zhu JG, Guo YG, Lin ZW, Li YJ, Huang YK (2011) Development of PM transverse flux motors with soft magnetic composite cores. IEEE Trans Magn 47(10):4376–4383
30. Guo YG, Zhu JG, Watterson PA, Wei Wu (2006) Development of a PM transverse flux motor with soft magnetic composite core. IEEE Trans Energy Conver 21(2):426–434
31. Lei G, Guo YG, Zhu JG et al (2012) System level six sigma robust optimization of a drive system with PM transverse flux machine. IEEE Trans Magn 48(2):923–926
32. Lei G, Zhu JG, Guo YG, Hu JF, Xu W, Shao KR (2013) Robust design optimization of PM-SMC motors for Six Sigma quality manufacturing. IEEE Trans Magn 49(7):3953–3956
33. Lei G, Guo YG, Zhu JG, Chen XM, Xu W (2012) Sequential subspace optimization method for electromagnetic devices design with orthogonal design technique. IEEE Trans Magn 48 (2):479–482
34. Lei G, Xu W, Hu JF, Zhu JG, Guo YG, Shao KR (2014) Multi-level design optimization of a FSPMM drive system by using sequential subspace optimization method, IEEE Trans Magn, 50(2), Article no 7016904
35. Liu CC, Zhu JG, Wang YH, Lei G, Guo YG, Liu XJ (2014) A low-cost permanent magnet synchronous motor with SMC and ferrite PM, In: Proceedings of 17th international conference on electrical Machines and Systems (ICEMS), pp 397–400
36. Fei W, Luk PCK, Shen JX, Wang Y, Jin M (2012) A novel permanent-magnet flux switching machine with an outer-rotor configuration for in-wheel light traction applications. IEEE Trans Ind Appl 48(5):1496–1506

37. Hua W, Cheng M, Zhu ZQ, Howe D (2006) Design of flux-switching permanent magnet machine considering the limitation of inverter and flux-weakening capability. In: proceedings of 41st IAS annual meeting-industry applications conference, vol 5, pp 2403–2410
38. Morio J (2011) Global and local sensitivity analysis methods for a physical system. Eur J Phys 32(6):1577–1583
39. Rodriguez-Fernandez M, Banga JR, Doyle FJ (2012) Novel global sensitivity analysis methodology accounting for the crucial role of the distribution of input parameters: application to systems biology models. Int J Robust Nonlin Control 22:1082–1102
40. Morris MD (1991) Factorial sampling plans for preliminary computational experiments. Technometrics 33(2):161–174
41. Sobol IM (2011) Global sensitivity indices for nonlinear mathematical models and their Monte Carlo estimates. Math Comput Simul 55:271–280
42. Herman JD, Kollat JB, Reed PM, Wagener T (2013) Technical note: method of morris effectively reduces the computational demands of global sensitivity analysis for distributed watershed models. Hydrol Earth Syst Sci 17:2893–2903
43. Li QS, Liu DK, Leung AYT, Zhang N, Luo QZ (2002) A multi-level genetic algorithm for the optimum design of structural control systems. Int J Numer Meth. Engng 55:817–834
44. Wang SH, Meng XJ, Guo NN, Li HB, Qiu J, Zhu JG et al (2009) Multi-level optimization for surface mounted PM machine incorporating with FEM. IEEE Trans Magn 45(10):4700–4703
45. Meng XJ, Wang SH, Qiu J, Zhu JG, Wang Y, Guo YG et al (2010) Dynamic multi-level optimization of machine design and control parameters based on correlation analysis. IEEE Trans Magn 46(8):2779–2782
46. Meng XJ, Wang SH, Qiu J, Zhang QH, Zhu JG, Guo YG, Liu DK (2011) Robust multi-level optimization of PMSM using design for six sigma. IEEE Trans Magn 47(10):3248–3251
47. Lei G,Liu CC, Guo YG, Zhu JG (2015) Multi-disciplinary design analysis for PM motors with soft magnetic composite cores, IEEE Trans Magn, 51(11), Article no 8109704
48. Huang YK, Zhu JG et al (2009) Thermal analysis of high-speed SMC motor based on thermal network and 3D FEA with rotational core loss included. IEEE Trans Magn 45(106):4680–4683

Chapter 5
Design Optimization Methods for Electrical Drive Systems

Abstract Electrical drive systems are key components in modern appliances, industry equipment and systems, such as digital machine tools and hybrid and pure electric vehicles. To obtain the best performance of these drive systems, the motors and their control systems should be designed and optimized simultaneously at the system level rather than the component level. This chapter presents system-level design and optimization methods for electrical drive systems, namely the single-level optimization method, multi-level optimization method, and multi-level Genetic Algorithm (MLGA). Two electrical drive systems are investigated to illustrate the effectiveness of those proposed methods. The performances of two machines are evaluated by the finite element models, which have been verified by comparing with the experimental results on prototypes. The proposed multi-level method can increase the performance of the whole drive system, such as higher output power, lower material cost and lower dynamic overshoot, and decrease the computational cost significantly compared with those of single-level design optimization method.

Keywords Electrical drive systems · System-level design optimization · Multi-level design optimization · Field oriented control · Finite element methods · Model predictive control · Transverse flux machine · Permanent magnet synchronous machine

5.1 Introduction

Electrical machines and the corresponding drive systems have a history of over a century and the design procedure has become almost "standard". When designing an appliance that needs an electrical drive system, the designer firstly selects the motor, inverter/converter and controller from the existing products. The appliance designer, on one hand, has to deliver the functions that the appliance is supposed to have, and on the other hand, has to take into account the availability and performance that the existing motor drive can provide. This motor manufacturer-oriented approach has been the dominant design concept for drive systems for a long time.

G. Lei et al., *Multidisciplinary Design Optimization Methods for Electrical Machines and Drive Systems*, Power Systems,
DOI 10.1007/978-3-662-49271-0_5

However, this approach would apply many constraints to the design and therefore limit the functions of the appliance [1].

With the fast development of CAD/CAE software, new material, flexible mechanical manufacturing technology, advanced optimization and control algorithms, it is possible to design a motor to meet the special requirements of a particular application. Since early 1990s, this application oriented approach has become a common practice. Nowadays, the motors and their control systems are generally closely integrated into the appliances. Therefore, more and more holistic integrated design problems of the electrical drive systems have boomed in industry, for example, the drive systems for hybrid electric vehicles (HEVs) [2–4].

Through the extensive research practice, it is recognized that when designing such an electrical drive system, it is important to pursue the optimal system performance rather than the optimal components like motor, because assembling individually optimized components into a system cannot necessarily guarantee an optimal system performance. The optimal system performance can only be achieved through a holistic approach of integrated simultaneous optimization of all components at the system level [1, 5, 6].

Figure 2.1 illustrated a general design framework and the interactions between different disciplines/domains for electrical drive systems. As shown, the design optimization of such a drive system is really a multi-disciplinary problem. It mainly includes electromagnetic, material, mechanical, thermal and power electronic designs, which are strongly coupled [7–9]. In order to achieve high system-level performance, the perfect cooperation of motor and its drive and control systems must be designed and optimized simultaneously.

Although the importance of system-level design optimization of electrical drive systems is noted, not much work has been reported in the literature [1, 5, 6]. Traditional design and optimization methods are mostly on the component level of different kinds of motors [10–16]. Generally, cogging torque, torque ripple, cost, weight and energy consumption are the main concerns for motors' performance parameters in the design and optimization process [17–19]. For the design optimization of these motors, Chap. 4 presented several kinds of optimization methods, including the combinations of intelligent optimization algorithms and finite element model (FEM) or approximate models, for example, the differential evolution algorithm (DEA) plus Kriging model. Approximate models are generally used to replace the FEM in the performance evaluation of motors so as to reduce the FEM computational costs [1, 5, 10, 20, 21].

On the other hand, for the controller part, though a lot of control algorithms have been developed, such as field oriented control (FOC), direct torque control (DTC) and model predictive control (MPC) [22–26], they are also generally designed and optimized on the controller level, and have not been combined with the design optimization of motors [27].

This component-level-based method may be reasonable for some traditional motors and their drive systems where there is much design experience that can be used. However, there is not much design experience for novel electrical drive systems. Furthermore, these methods are not system-level holistic design basically.

As previously discussed, by this component level approach, one can hardly achieve the optimal system performance. Therefore, how to design and optimize novel high performance drive systems is an important problem in both research community and industrial applications [1, 5].

In order to deal with the above problems, this chapter presents three types of system-level design optimization methods for electrical drive systems. This chapter is structured as follows. Section 5.2 presents the system-level design optimization framework and models for electrical drive systems. Section 5.3 presents a single-level (only at the system level) optimization method for the design of drive systems. Section 5.4 presents a multi-level design optimization method for drive systems, including the investigation of the first design example: a drive system consists of a permanent magnet (PM) transverse flux machine (TFM) with soft magnetic composite (SMC) core and an improved MPC system. Section 5.5 introduces the multi-level Genetic Algorithm (MLGA) for the design optimization of drive systems, including the second drive system example which is composed of a surface-mounted permanent magnet synchronous machine (SPMSM) and a classical FOC system, followed by the summary section.

5.2 System-Level Design Optimization Framework

Figure 2.1 briefly illustrated a multi-disciplinary (or multi-domain) design framework for electrical drive systems. However, the design modules are strongly coupled and it is not easy to derive the design and optimization flowchart from this framework. Figure 5.1 shows a deductive system-level design and optimization framework for electrical drive systems [1]. It mainly includes five steps, namely system inputs, selection, design, optimization, and evaluation, as the following:

Step 1 Determination of system's requirements and specifications. In this step, system's design objectives and constraints, such as cost, weight, torque ripple and motor efficiency, have to be considered and defined.

Step 2 Selection or design of motor type, drive and controller type with respect to the system specifications. This step can be done through a qualitative comparison based on literature survey and experience. A drive system in general consists of two parts, namely motor and controller, and the latter includes a power electronic converter and drive control algorithm. There are some interactions between these two parts, e.g. a special type of controller fits the given type of motor better than the others.

Step 3 Design of motor and controller jointly. The motor design consists of material selection and modelling, electromagnetic and thermal designs, and so on [7–9]. The controller design mainly includes the design of control algorithms and parameters. These two designs are done simultaneously.

Step 4 Construction of design optimization models for the motor, controller and the whole system. The motor design optimization model can be defined as

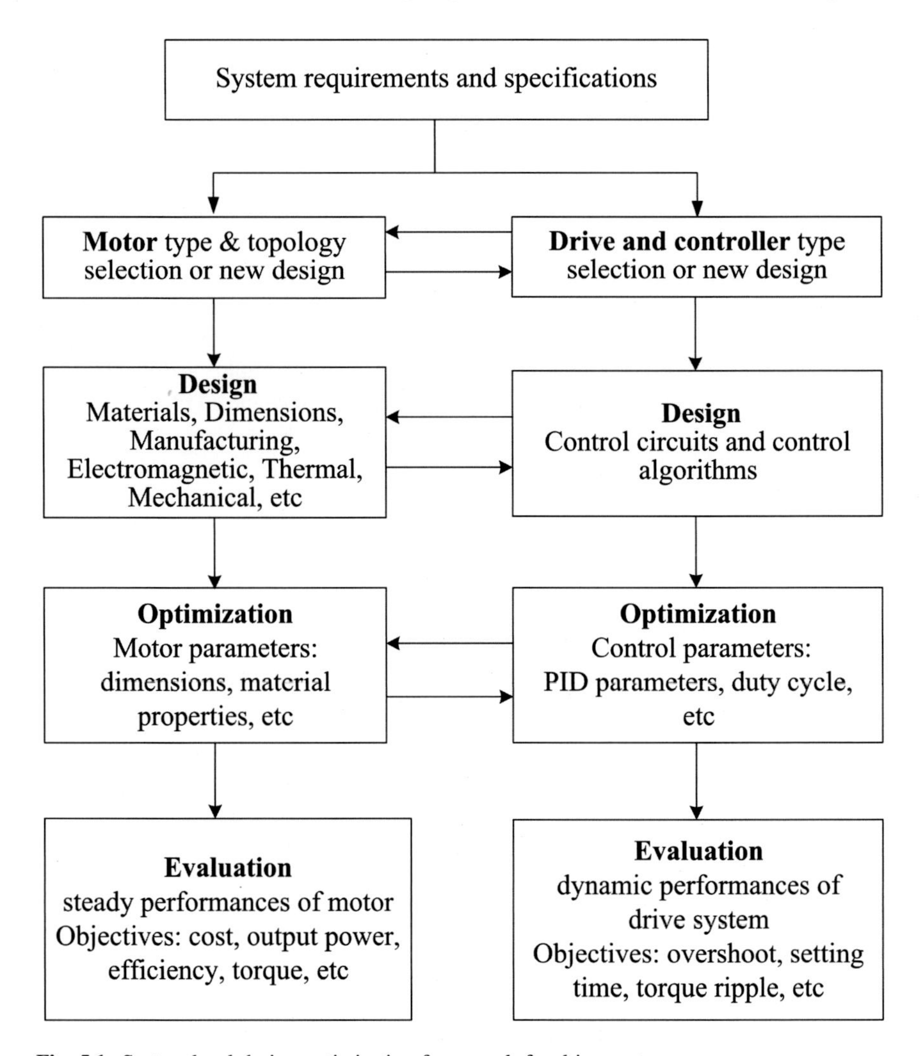

Fig. 5.1 System-level design optimization framework for drive systems

$$\begin{array}{ll} \min: & f_m(\mathbf{x}_m) \\ \text{s.t.} & g_{mi}(\mathbf{x}_m) \leq 0,\ i = 1, \ldots, N_m, \\ & \mathbf{x}_{ml} \leq \mathbf{x}_m \leq \mathbf{x}_{mu} \end{array} \tag{5.1}$$

where $\mathbf{x}_m$, f_m and g_m are the motor design parameter vector, objectives and constraints, $\mathbf{x}_{ml}$ and $\mathbf{x}_{mu}$ the lower and upper boundaries of $\mathbf{x}_m$, respectively, and N_m is the number of the constraints. It should be noted that the objectives and constraints in (5.1) must be defined in terms of the required system objectives and constraints in Step 1.

The control design optimization model can be defined as

$$\begin{array}{ll} \min: & f_c(\mathbf{x}_c) \\ \text{s.t.} & g_{ci}(\mathbf{x}_c) \leq 0,\ i = 1,\ldots,N_c\,, \\ & \mathbf{x}_{cl} \leq \mathbf{x}_c \leq \mathbf{x}_{cu} \end{array} \tag{5.2}$$

where $\mathbf{x}_c$, f_c and g_c are the control design parameter vector, objectives and constraints, $\mathbf{x}_{cl}$ and $\mathbf{x}_{cu}$ the lower and upper boundaries of $\mathbf{x}_c$, respectively, and N_c is the number of controller constraints. Similarly, the objectives and constraints in (5.2) must also be defined in terms of the required system objectives and constraints.

Combining the motor and controller design optimization models, (5.1) and (5.2), one obtains the system-level design optimization model as the following:

$$\begin{array}{ll} \min: & f_s(\mathbf{x}_s) = F(f_m, f_c) \\ \text{s.t.} & g_{mi}(\mathbf{x}_s) \leq 0,\ i = 1,2,\ldots,N_m \\ & g_{ci}(\mathbf{x}_s) \leq 0,\ i = 1,2,\ldots,N_c \\ & \mathbf{x}_{sl} \leq \mathbf{x}_s \leq \mathbf{x}_{su} \end{array}, \tag{5.3}$$

where $\mathbf{x}_s = [\mathbf{x}_m, \mathbf{x}_c]$, $\mathbf{x}_{sl}$ and $\mathbf{x}_{su}$ are the lower and upper boundaries of $\mathbf{x}_s$, respectively, and f_s is the system objective which is generally a function of f_m and f_c.

Step 5 Evaluation of the system performance. This step consists of two parts. One is the evaluation of steady performance of motor, such as cost and efficiency. The other is the evaluation of dynamic performance of controller or the whole drive system, such as overshoot, settling time, torque ripple, and speed ripple.

5.3 Single-Level Design Optimization Method

Figure 5.2 illustrates the first type of optimization method for electrical drive systems. It can be seen that the optimization process is implemented at a single level for the whole system, which is thus known as the single-level design optimization method [1].

This method mainly includes the following three steps:

Step 1 Determination of system level optimization model (5.3). It includes the selection of motor and controller for the specific drive system.

Step 2 Selection of an optimization method. As drive systems are always high dimensional and non-linear design problems, intelligent algorithms, such as genetic algorithm (GA) and DEA, can be good choices in many situations. Therefore, the algorithm parameters should be determined in this step, such as genetic operators in GA and mutation operator in DEA.

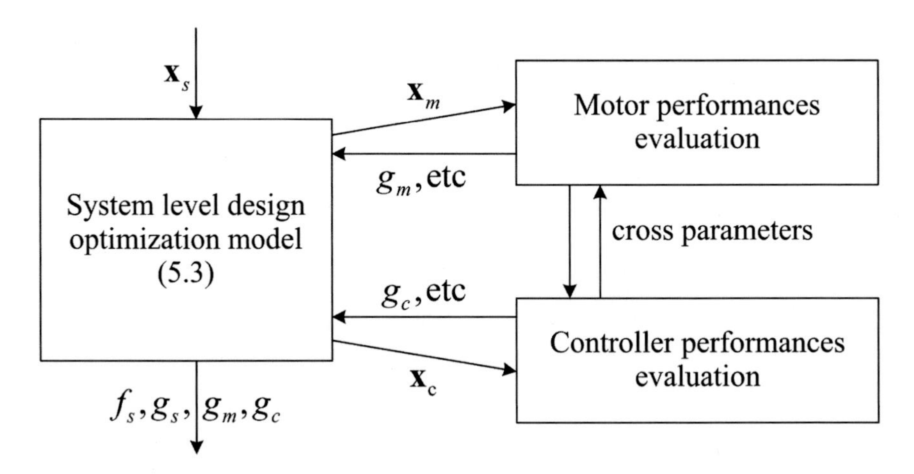

Fig. 5.2 Single-level optimization method for electrical drive systems

Step 3 Implementation of optimization process. Firstly, generate initial population of $\mathbf{x}_s$. Secondly, evaluate the drive system performance parameters, objectives, and constraints in (5.3). Thirdly, implement the optimization algorithm until the convergence criteria are met. Finally, terminate and output the optimal solutions.

However, the computational cost of this single-level optimization method is always very high as these design problems are generally high dimensional, non-linear, and strongly coupled multi-domain design analyses. As different domains have different analysis techniques and software, the computational cost of whole system is very expensive. For example, power electronic circuit analysis is needed in the control design, but the needed characteristic parameters of motor are generally calculated by FEM in the motor design, i.e. the power electronic circuit design and electromagnetic design are strongly coupled in electrical drive systems. The computational cost of finite element analysis is usually very expensive in most cases, especially for some motors of complex structures. To overcome these problems, a multi-level design optimization method is presented as follows [1].

5.4 Multi-level Design Optimization Method

5.4.1 Method Flowchart

Figure 5.3 depicts a multi-level design optimization framework for electrical drive systems. Three levels are considered in this framework, namely the motor, control, and system levels [1].

This optimization method includes three steps as follows:

Step 1 Determination of optimization models (5.1) and (5.2) for the motor and control levels, respectively. All the required system objectives and constraints should be defined in (5.1) and (5.2), so that only two levels, the motor and control levels, are needed to be optimized in this framework.

Step 2 Optimization. This step includes the optimization processes for motor and control levels, respectively.

The motor level—The aim of this level is to optimize the motor model (5.1) and evaluate the motor steady state performance, such as the cost, weight, output power and efficiency. The motor characteristic parameters should be calculated in this step, such as the winding resistance, inductance and magnetic flux for the design optimization of the control level.

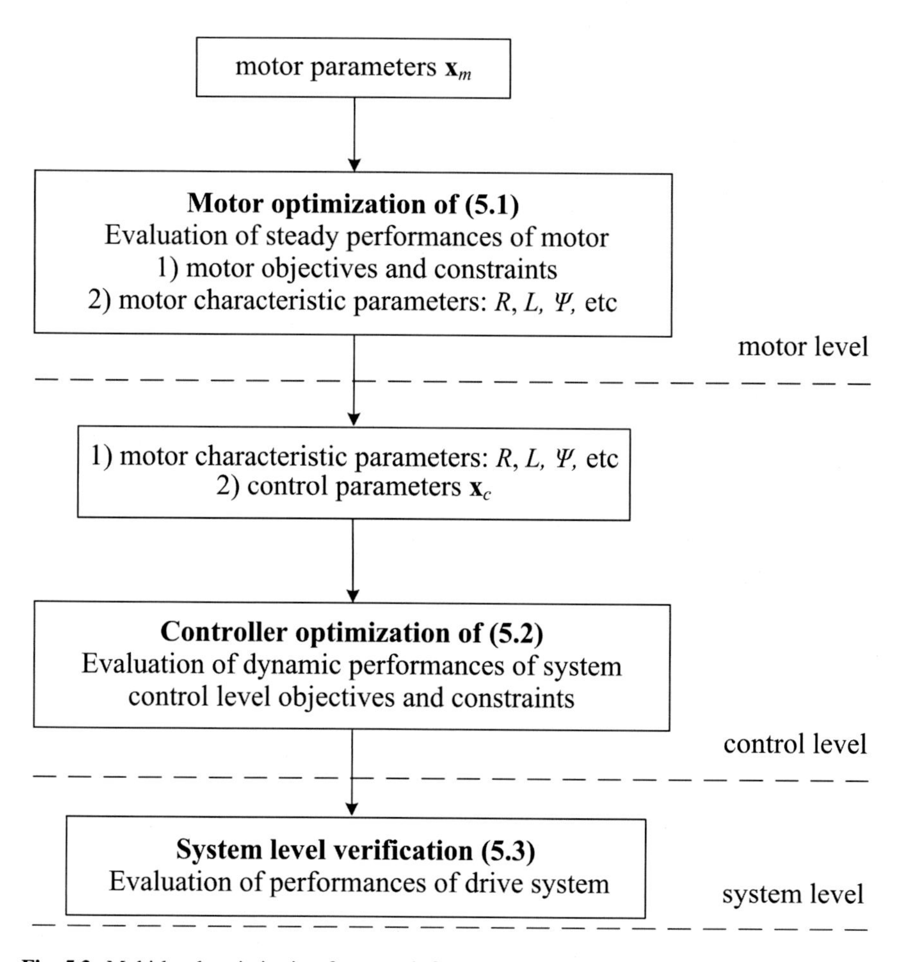

Fig. 5.3 Multi-level optimization framework for electrical drive systems

The control level—The aim of this level is to optimize the control model (5.2) and evaluate the system dynamic performance, such as the overshoot and settling time.

Step 3 Verification of system level performance (5.3). The aim of this step is to evaluate the system performance and output the optimization results.

As outlined above, the drive system design optimization is always high dimensional and nonlinear. The optimization efficiency of multi-level optimization method mainly depends on two issues. The first one is how to construct an efficient multi-level optimization framework, especially for high dimensional problems. The second one is how to reduce the computational cost of optimization models, which are the main contents of the previous chapter. Therefore, the proposed new optimization methods for electrical machines can be employed for the optimization of motor and control levels, respectively. For example, if there are seven optimization parameters for the motor, then the multi-level optimization method presented in Sects. 4.6 and 4.7 can be introduced to improve the optimization efficiency, and the sensitivity analysis techniques presented in Sect. 4.5 can be employed to determine the multi-level optimization framework.

5.4.2 *Design Example for a Drive System of TFM and MPC*

5.4.2.1 Design Optimization Model for Motor Level

In this example, we will investigate a drive system consisting of a PM-SMC TFM and an improved MPC control system. More details of this motor can be found in the previous chapters.

Figure 5.4 shows the structure of one phase of this TFM with main dimensions shown in Table 2.1. Figure 5.5 shows the FEA model for this TFM, which contains one pole pitch of a phase because of the symmetry. The main design and optimization parameters are also shown in this figure.

To optimize this machine, eight parameters are considered as the optimization variables as shown in Table 5.1 (more motor dimensions can be seen in Table 2.1) and Fig. 5.5. All these parameters should be optimized to minimize the cost of material and maximize the output power of the motor. The objective cost mainly includes the material costs of PMs, copper, SMC core and steel. Four constraints are also considered for this machine. The optimization model can be defined as follows:

$$\begin{aligned}
\text{min}: \quad & f_m(\mathbf{x}_m) = w_1 \frac{Cost}{C_0} + w_2 \frac{P_0}{P_{\text{out}}} \\
\text{s.t.} \quad & g_{m1}(\mathbf{x}_m) = 0.795 - \eta \leq 0, \\
& g_{m2}(\mathbf{x}_m) = 640 - P_{\text{out}} \leq 0, \\
& g_{m3}(\mathbf{x}_m) = sf - 0.8 \leq 0, \\
& g_{m4}(\mathbf{x}_m) = J_c - 6 \leq 0, \\
& \mathbf{x}_{ml} \leq \mathbf{x}_m \leq \mathbf{x}_{mu}
\end{aligned} \tag{5.4}$$

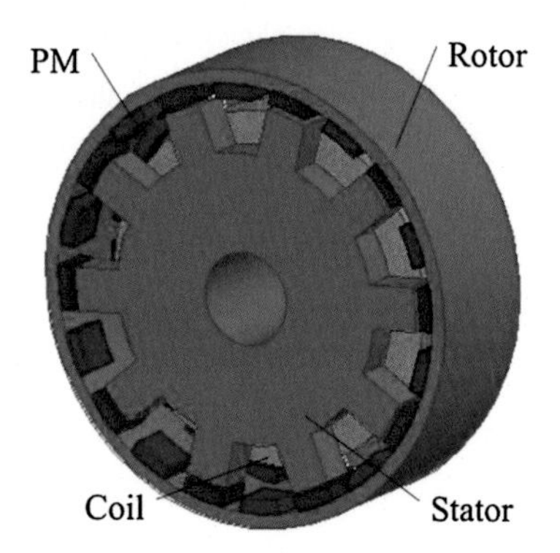

Fig. 5.4 Structure of the PM-SMC TFM with SMC stator (one phase)

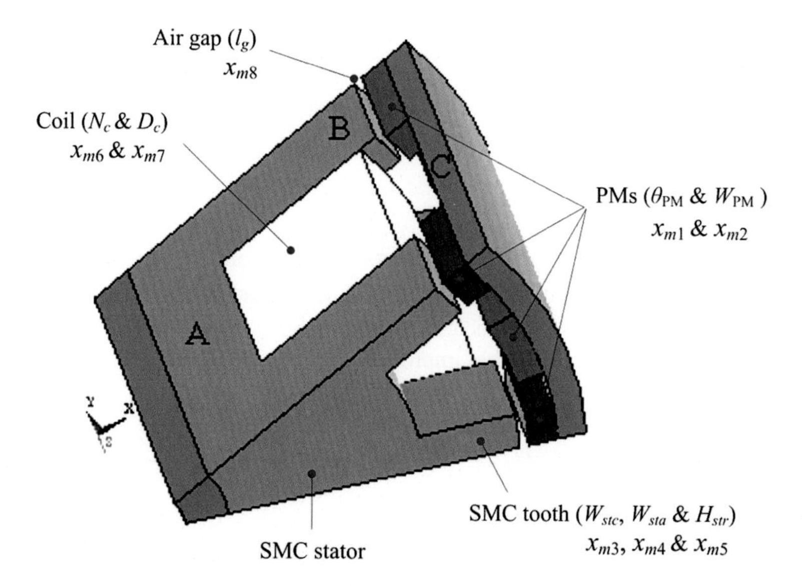

Fig. 5.5 FEM and optimization parameters for PM-SMC TFM

where w_1 and w_2 are weighting factors, C_0 and P_0 the cost and output power (P_{out}) of the initial PM TFM, η, sf and J_c the motor efficiency, winding fill factor, and current density, respectively [1, 5, 28, 29].

5.4.2.2 Design Optimization Model for Control Level

In Chap. 2, an improved MPC scheme with a duty ratio optimization module was presented to drive PM motors [30]. It will be used in this chapter as the control method for the PM-SMC TFM. This control scheme was shown in Fig. 2.37.

Table 5.1 Main motor design and optimization parameters

Par.		Description	Unit	Value	Min.	Max.	Step size
x_{m1}	θ_{PM}	PM circumferential angle	Deg.	12	9	12	0.05
x_{m2}	W_{PM}	PM width	mm	9	6	9	0.05
x_{m3}	W_{stc}	SMC tooth circumferential width	mm	9	8	10	0.05
x_{m4}	W_{sta}	SMC tooth axial width	mm	8	7	9	0.05
x_{m5}	H_{str}	SMC tooth radial height	mm	10.5	9	11	0.05
x_{m6}	N_c	Number of turns	–	125	110	128	1
x_{m7}	D_c	Diameter of copper wire	mm	1.25	1.0	1.3	0.01
x_{m8}	l_g	Air gap length	mm	1.0	0.95	1.15	0.01

As mentioned in Chap. 2, the key issue of MPC is the definition of cost function. Since the two greatest concerns of a PM motor are the torque and stator flux, the cost function is defined to ensure that both the torque and stator flux at the end of control period are as close as possible to the reference values. To illustrate the optimization parameters of the control level, the cost function is rewritten as

$$\begin{aligned} G = & |T_e^* - T_e^{k+1}| + k_1 \big| |\psi_s^*| - |\psi_s^{k+1}| \big| \\ & + A\left(|T_e^* - T_e^{k+N}| + k_1 \big| |\psi_s^*| - |\psi_s^{k+N}| \big| \right) \end{aligned} \tag{5.5}$$

where T_e^*, ψ_s^*, T_e^{k+1} and ψ_s^{k+1} are the reference torque and flux, predicted torque and flux, T_e^{k+N} and ψ_s^{k+N} the linear predictions of torque and flux at the $(k + N)$-th instant, and k_1 and A weighting factors, respectively.

The other important part of this improved MPC is the duty ratio module. The expression of duty ratio optimization module has the form as

$$d = \left| \frac{T_e^* - T_e^{k+1}}{C_T} \right| + \left| \frac{\psi_s^* - \psi_e^{k+1}}{C_\psi} \right|, \tag{5.6}$$

where C_T and C_ψ are two positive parameters. The idea of this method is that a larger difference between the reference and predicted torque values would lead to a larger duty ratio value [30].

Six parameters should be optimized in the control level. They are A, N, C_T, C_ψ, K_p and K_i, where K_p and K_i are the PI controller parameters as shown in Fig. 2.37. One objective and four constraints are considered for this level. The objective is to minimize the sum of root mean square errors (RMSE) of torque (T) and speed (n) in the steady state operation period. At the same time, the speed overshoot should be minimized for this control system. The optimization model of the control system can be expressed as

$$
\begin{aligned}
\text{min}: \quad & f_c(\mathbf{x}_c) = w_3 \frac{\text{RMSE}(T)}{T_{rated}} + w_4 \frac{\text{RMSE}(n)}{n_{rated}} + w_5 n_{os} \\
\text{s.t.} \quad & g_{c1}(\mathbf{x}_c) = \text{RMSE}(T)/T_{rated} - 0.06 \leq 0, \\
& g_{c2}(\mathbf{x}_c) = \text{RMSE}(n)/n_{rated} - 0.05 \leq 0, \\
& g_{c3}(\mathbf{x}_c) = n_{os} - 0.02 \leq 0, \\
& g_{c4}(\mathbf{x}_c) = t_s - 0.02 \leq 0, \\
& \mathbf{x}_{cl} \leq \mathbf{x}_c \leq \mathbf{x}_{cu}
\end{aligned} \tag{5.7}
$$

where w_3 to w_5 are weighting factors, the subscript *rated* indicates that the values are obtained from the motor optimization model (5.4), n_{os} is the speed overshoot, which should be no larger than 2 % of the rated speed, 1800 rev/min, and t_s the settling time, which should be no larger than 0.02 s after the load is applied to the control system [1].

5.4.2.3 Optimization Flowchart and Results

A.Multi-level Optimization Flowchart

Firstly, for the eight parameters at motor level, it was found that they can be divided into two subspaces according to our design experience [1]. The first subspace X1 includes x_{m1}, x_{m2}, x_{m6} and x_{m7}, which are significant to the cost and output power of the motor. The second subspace X2 includes x_{m3}, x_{m4}, x_{m5} and x_{m8}. Therefore, the optimization flowchart of the motor level has two sublevels.

Secondly, for the six parameters in the control level, after the Design of experiments (DOE) analysis, it is found that except the third control parameter C_T, the other parameters have the same significant level. Table 5.2 shows the analysis of variance (ANOVA) results for the control level. As shown, the second column means the sum of square of deviations, the third column DF the degree of freedom, the fourth column Var. the variance, and the F column the value of hypothesis testing with F distribution. F_α is a reference value, and α is significant level for hypothesis test. 0.01 is a generally used value for α. If the value of F is larger than F_α, the corresponding factor is a significant factor. Therefore, only the third parameter is a significant factor for the objective of control level. Theoretically, we can use multi-level optimization method with two subspaces for the control level.

Table 5.2 DOE and ANOVA data for control level

Source	Sum.	DF	Var.	F	F_α	Sig.
x_{c1}	0.013	4	0.003	0.80	2.87	
x_{c2}	0.015	4	0.004	0.88	2.87	
x_{c3}	0.057	4	0.014	3.44	2.87	*
x_{c4}	0.020	4	0.005	1.24	2.87	
x_{c5}	0.013	4	0.003	0.76	2.87	
x_{c6}	0.022	4	0.006	1.33	2.87	
Error	0.083	20	0.004	–	–	–
Total	0.139	24	–	–	–	–

However, since the significant space only has 1 factor, we just select all six parameters as the same level.

In summary, the total optimization framework of this drive system has three levels as shown in Fig. 5.6. The first and second levels are the subspaces X1 and X2, respectively, which should be optimized with model (5.4). After the optimization of motor level, the motor characteristic parameters, such as R, L and flux can be obtained. They will be used as the input parameters of MPC control system in the control level. The third level is the subspace of all the control parameters in model (5.7).

B.Optimization Results

First of all, DEA is selected as the optimization algorithm in the multi-level optimization of this drive system. The algorithm parameters include the mutation scaling factor of 0.8, crossover factor of 0.8, and the maximum number of iterations of 1000. Then ε in the multi-level optimization method is defined as 1 %. All weighting factors are assumed to be 1 in this work. Tables 5.3 and 5.4 list the optimization results of motor and control levels obtained by the single-level and multi-level optimization methods, respectively. From these tables, the following conclusions can be drawn:

(1) The motor level. For the initial design scheme, the motor efficiency is 79.5 %, the output power 640 W, average torque 3.40 Nm, and material cost $35.8. By the single-level optimization method with DEA and FEM, all the 14 parameters (8 motor and 6 control parameters) are optimized as shown in Fig. 5.2. The obtained motor efficiency is 81.5 %, output power 658 W, average torque 3.49 Nm, and material cost $28.3. They are better than those of the initial design.
For the multi-level optimization method, 3 iteration processes are required to get the optimal results. Figure 5.7 shows the iteration process of multi-level optimization for the motor level. As shown, level 1 is optimized twice while level 2 is optimized only once. After the optimization, the optimal output power reaches 670 W and the average torque 3.55 Nm while the motor efficiency decreases to 81.3 %. The cost is only $26.9, which is the smallest among these three designs, and the output power is increased by 4.7 % (30/640) by using the proposed multi-level optimization method compared with the initial design.
(2) The control level. As shown in Table 5.4, the relative RMSEs of torque and speed are 4.17 and 0.10 %, respectively, the speed overshoot is 1.03 % and the objective 5.30 % for control level by using the single-level optimization method. After the multi-level optimization, all these objectives have been increased significantly. For example, the relative RMSE of speed has been

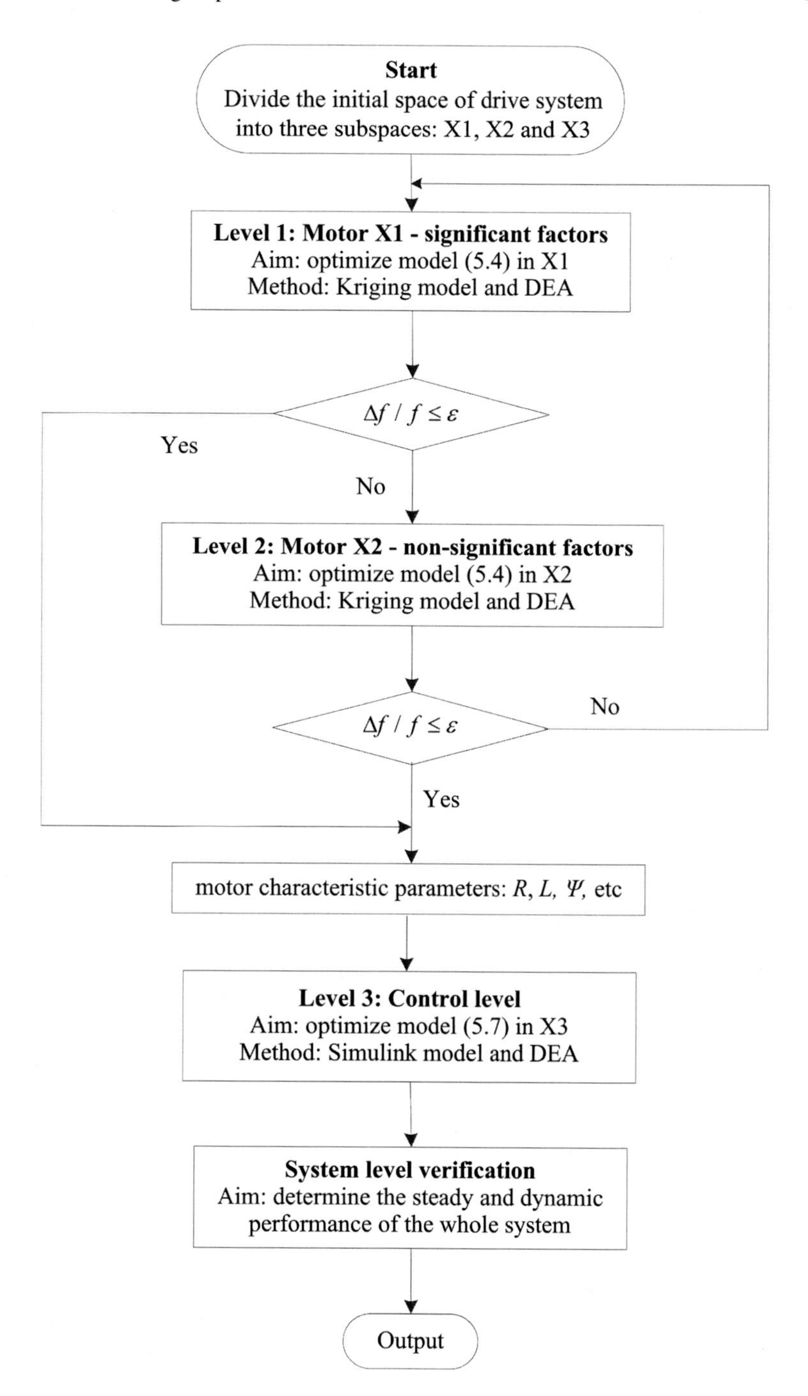

Fig. 5.6 Multi-level optimization flowchart for PM TFM drive system

Table 5.3 Optimization results of TFM parameters (Motor level)

Par.	Unit	Initial	Single-level	Multi-level
x_{m1}	Deg.	12	10.65	10.00
x_{m2}	mm	9	8.00	7.65
x_{m3}	mm	9	8.45	8.0
x_{m4}	mm	8	7.65	7.95
x_{m5}	mm	10.5	9.15	10.9
x_{m6}	turn	125	117	110
x_{m7}	mm	1.25	1.23	1.27
x_{m8}	mm	1.0	1.00	0.95
η	%	79.5	81.5	81.3
P_{out}	W	640	658	670
sf	–	0.56	0.55	0.50
J_c	A/mm^2	4.72	5.88	5.96
T	Nm	3.40	3.49	3.55
$Cost$	$	35.8	28.3	26.9
f_m	–	–	1.70	1.65
FEM	–	–	14,000	16,25

Table 5.4 Optimization results of MPC parameters (control level)

Par.	Single-level	Multi-level
x_{c1}	0.320	0.386
x_{c2}	6	7
x_{c3}	0.959	1.17
x_{c4}	0.03427	0.03568
x_{c5}	0.199	0.23
x_{c6}	2.698	1.619
RMSE(T)/T_{rated}	4.17 %	3.95 %
RMSE(n)/n_{rated}	0.10 %	0.02 %
n_{os}	1.03 %	0.90 %
t_s	0.01	0.01
f_c	5.30 %	4.87 %
Simulation calls	14,000	6,000

decreased from 0.10 to 0.02 %, and the objective of control level from 5.30 to 4.87 %. Therefore, the dynamic performances have been greatly improved by using the multi-level optimization method. Figure 5.8 illustrates the dynamic performance of the drive system obtained by the multi-level method. As shown, the dynamic performances of speed and torque are very good.

(3) For the computational cost, the cost of FEM analysis at the motor level and the cost of Simulink simulation calls at control level are the largest computational burden for the whole optimization process. For the single-level optimization method with DEA, about 14,000 FEM samples and 14,000 Simulink simulation calls (5 $\times$14 $\times$ 200, where 5 $\times$ 14 is the population size and 200 the

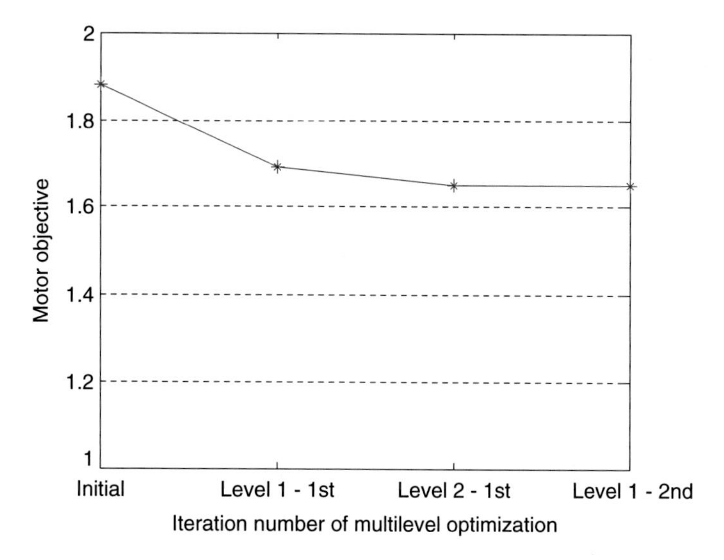

Fig. 5.7 Iteration process of multi-level optimization for the motor level

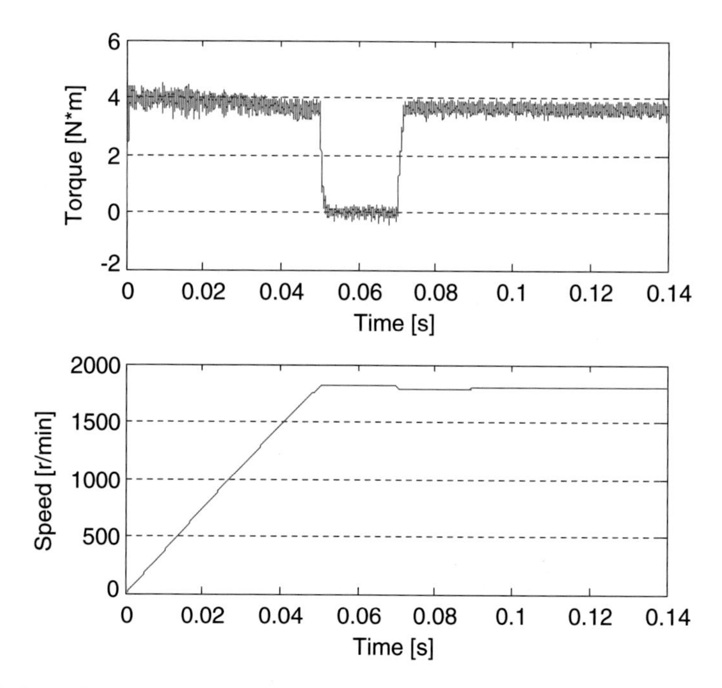

Fig. 5.8 Dynamic performance of TFM with optimized MPC parameters

average iteration number of DEA) are needed to achieve the optimal results. Actually, it is hard for an intelligent algorithm to deal with this kind of high-dimensional and highly nonlinear optimization problem. It is time consuming and tends to find a local optimal point.

On the other hand, only 1625 (see Table 5.3) FEM samples are needed for the motor level optimization by using the proposed multi-level optimization method, which is about 11.6 % of the direct single-level optimization method. Moreover, about 6,000 Simulink simulation calls are needed for the control level by using the multi-level optimization method. This is less than half of the simulation cost of single level optimization method. Therefore, the proposed multi-level design optimization method can significantly reduce the computational cost, and produce better solutions than the single-level optimization method.

In summary, compared with the schemes gained from the initial design and single-level method, the solutions obtained by the multi-level optimization method have many improvements, such as larger output power, less material cost, and less overshoot. As a matter of fact, both the steady state and dynamic performances of this drive system have been improved by using the multi-level method.

5.5 MLGA for a SPMSM Drive System with FOC

5.5.1 Optimization Model

MLGA was introduced as a kind of optimization method for the high dimensional optimization problems of electrical machines in Chap. 4. Its efficiency has been verified by a design example of an SPMSM [31, 32]. MLGA can be applied to design optimization of electrical drive systems with a similar optimization framework. In this section, MLGA is presented for design optimization of a motor drive system consisting of a SPMSM and FOC control scheme. The main design parameters of this SPMSM can be seen in Sect. 4.7, and the FOC control scheme in Fig. 2.22. As shown, three PI controllers are used for the d- and q-axis components of stator current, and speed control, respectively.

For the motor level optimization, the objective is to minimize the cost of copper and permanent magnets, and to maximize the motor efficiency, η. The optimization model can be expressed as

$$\begin{aligned} \max: \quad & f_m(\mathbf{x}_m) = \frac{K}{w_1 \frac{100-\eta}{100} + w_2 \frac{Cost}{Cost_{\max}}} \\ \text{s.t.} \quad & P_2 > 945W \\ & sf < 78\,\% \end{aligned} \tag{5.8}$$

where the design variable $\mathbf{x}_m = [hm, bm, Ns, WindD]$, Ns and $WindD$ are discrete variables, K, w_1 and w_2 the weighting factors defined by the designer, P_2 is the output power, and sf the fill factor [6].

For the control level optimization, the six integral and proportional gain factors in the three PI controllers as shown in Fig. 2.22 are chosen as the design optimization variables. The optimization model can be defined as

$$\begin{array}{ll} \min: & f_c(\mathbf{x}_c) = \alpha_1 T_{rip} + \alpha_2 n_{os} + \alpha_3 I_d \\ \text{s.t.} & T_{rip} \leq 0.5 \text{ Nm} \\ & n_{os} \leq 0.5\ \% \\ & I_d \leq 0.45 \text{ A} \end{array} \tag{5.9}$$

where $\mathbf{x}_c$ stands for the vector of six PI variables, T_{rip} is the torque ripple, n_{os} the overshoot of speed, I_d the d-axis component of stator current, and α_i (i = 1, 2, 3) are weighting factors.

5.5.2 *Optimization Framework*

As mentioned in Chap. 4, the problem matrix is a method to determine the multi-level optimization framework for MLGA, which can be conducted by using the correlation analysis and DOE techniques. Note that the optimization model and parameters of the motor level of this drive system are the same as those investigated in the MLGA for SPMSM in Chap. 4. Therefore, the same multi-level optimization structure obtained in that chapter can be applied to the motor level of this drive system. That is, *hm* and *bm* are the variables of level 1 (indicated as X1), and *Ns* and *WindD* are the parameters in level 2 (indicated as X2). For the control level, all parameters can be placed into one level, i.e. level 3 (indicated as X3). Therefore, a three-level optimization framework as shown in Fig. 5.9 can be constructed for this drive system.

5.5.3 *Optimization Results*

In the implementation of MLGA, all weighting factors are defined as 1 in this work. Table 5.5 lists the optimization results obtained by MLGA, and Table 5.6 the proportional and integral gains calculated on the third level. It can be seen that the motor efficiency and output power have been increased greatly by using MLGA compared with the initial design. For example, the motor efficiency after MLGA optimization reaches 86.4 %, an increase by 2.7 % compared with the initial 83.7 %. Figure 5.10 illustrates the speed responses of SPMSM before and after the

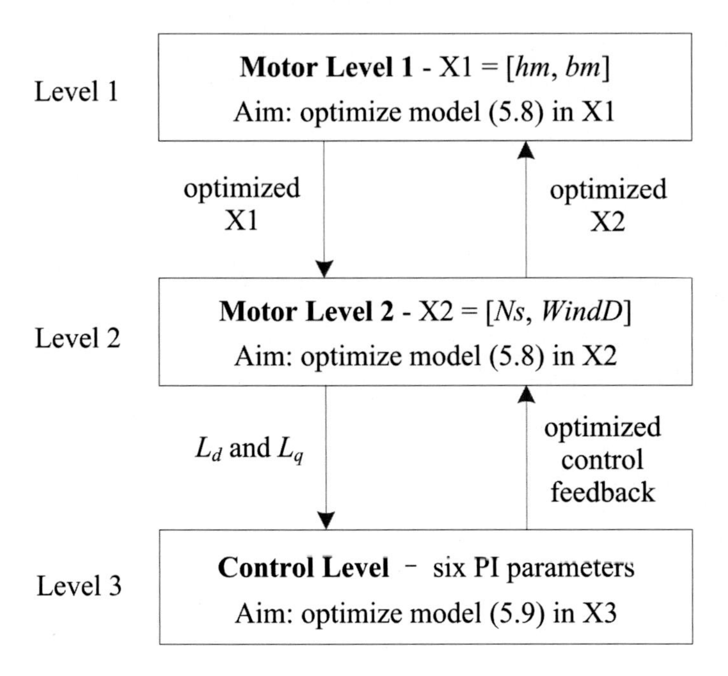

Fig. 5.9 Three-level optimization structure for a SPMSM drive system

Table 5.5 Optimal results for SPMSM on levels 1 and 2

Par.	Description	Unit	Initial	MLGA
hm	Thickness of PM	mm	1.8	2.3
bm	Width of PM	mm	31.4	30.3
Ns	Conductors per slot	turn	72	67
D	Diameter of conductor	mm	0.5	0.56
I_q	*q*-axis component of current	A	4.78	5.27
I_d	*d*-axis component of current	A	1.60	0.05
η	Efficiency	–	83.7 %	86.4 %
Cost	Cost of PM and winding	$	26.1	22.6
P_2	Output power	W	946	950
sf	Fill factor	%	67	78

optimization of PI controller parameters. The speed overshoot is about 0.7 rev/min, or 0.035 % of the rated speed (2000 rev/min) after the MLGA optimization in contrast to 6 rev/min, or 0.3 % of the rated speed, in initial design, a significant reduction of speed overshoot. In summary, the steady state and dynamic performances of the whole drive system have been improved greatly by using MLGA.

Table 5.6 Optimal results for control on level 3

Parameters	Initial	MLGA
Proportional gain in speed loop	1	18
Integral gain in speed loop	1	0.2
Proportional gain in I_d loop	1	20
Integral gain in I_d loop	1	0.32
Proportional gain in I_q loop	1	29
Integral gain in I_q loop	1	2

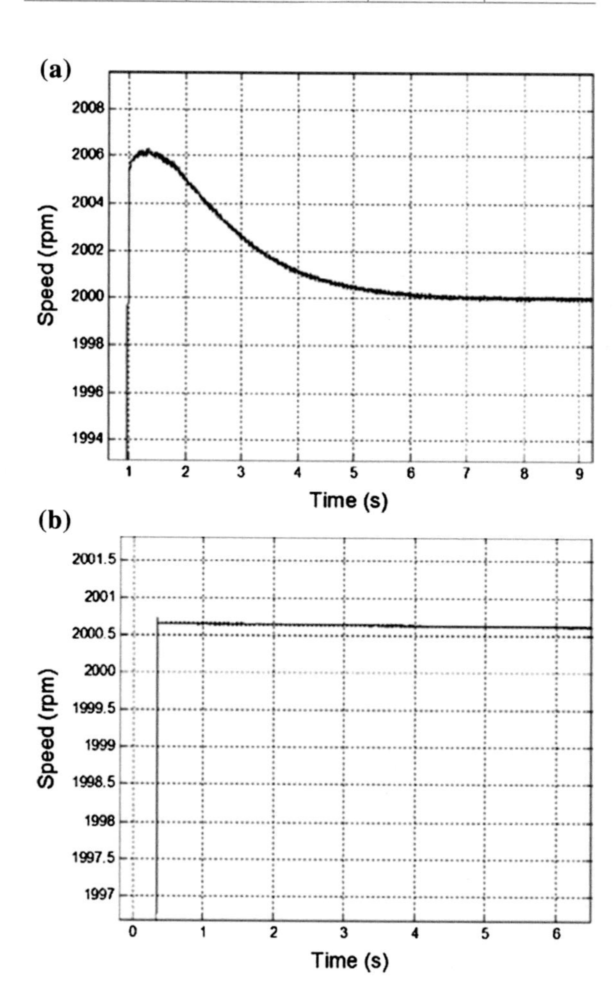

Fig. 5.10 Transient speed **a** before and **b** after optimization

5.6 Summary

From the above discussions, it can be seen that the system-level design optimization method is necessary for electrical drive systems in order to achieve high steady state and dynamic performances at the system level. The proposed multi-level method is efficient for the design optimization of high dimensional drive systems. This method will build a solid foundation to enable the effective development of novel high performance drive systems with new materials, low cost and high efficiency for industrial applications. This method can be also applied to other high dimensional design optimization problems in industrial applications. It will shorten the design cycle, reduce the design cost and improve the design efficiency for the industrial products in the early stage of product development.

References

1. Lei G, Wang TS, Guo YG, Zhu JG, Wang SH (2014) System level design optimization method for electrical drive system: deterministic approach. IEEE Trans Ind Electron 61 (12):6591–6602
2. Zhu ZQ, Howe D (2007) Electrical machines and drives for electric, hybrid, and fuel cell vehicles. Proc IEEE 95(4):746–765
3. Chan CC (2007) The state of the art of electric, hybrid, and fuel cell vehicles. Proc IEEE 95 (4):704–718
4. Emadi A, Lee YJ, Rajashekara K (2008) Power electronics and motor drives in electric, hybrid electric, and plug-in hybrid electric vehicles. IEEE Trans Ind Electron 55(6):2237–2245
5. Lei G, Guo YG, Zhu JG et al (2012) System level six sigma robust optimization of a drive system with PM transverse flux machine. IEEE Trans Magn 48(2):923–926
6. Meng XJ, Wang SH, Qiu J, Zhu JG, Wang Y, Guo YG et al (2010) Dynamic multi-level optimization of machine design and control parameters based on correlation analysis. IEEE Trans Magn 46(8):2779–2782
7. Lei G, Liu CC, Guo YG, Zhu JG (2015) Multi-disciplinary design analysis and optimization of a PM transverse flux machine with soft magnetic composite core," IEEE Trans Magn, 2015, vol 51(11). Article 8109704
8. Vese I, Marignetti F, Radulescu MM (2010) Multiphysics approach to numerical modeling of a permanent-magnet tubular linear motor. IEEE Trans Ind Electron 57(1):320–326
9. Kreuawan S, Gillon F, Brochet P (2008) Optimal design of permanent magnet motor using multi-disciplinary design optimisation. In: Proceedings of the 18th International Conference on Electrical Machines, 6–9 Sept., Vilamoura, pp 1–6
10. Lei G, Shao KR, Guo YG, Zhu JG (2012) Multi-objective sequential optimization method for the design of industrial electromagnetic devices. IEEE Trans Magn 48(11):4538–4541
11. Li WL, Zhang XC, Cheng SK, Cao JC (2013) Thermal optimization for a HSPMG used for distributed generation systems. IEEE Trans Ind Electron 60(2):474–482
12. Pfister P-D, Perriard Y (2010) Very-high-speed slotless permanent-magnet motors: analytical modeling, optimization, design, and torque measurement methods. IEEE Trans Ind Electron 57(1):296–303
13. Laskaris KI, Kladas AG (2011) Permanent-magnet shape optimization effects on synchronous motor performance. IEEE Trans Ind Electron 58(9):3776–3783
14. Yamazaki K, Ishigami H (2010) Rotor-shape optimization of interior-permanent-magnet motors to reduce harmonic iron losses. IEEE Trans Ind Electron 57(1):61–69
15. Ma C, Qiu L (2015) Multi-objective optimization of switched reluctance motors based on design of experiments and particle swarm optimization. IEEE Trans Energy Convers 30(3):1144–1153
16. Barcaro M, Bianchi N, Magnussen F (2012) Permanent-magnet optimization in permanent-magnet-assisted synchronous reluctance motor for a wide constant-power speed range. IEEE Trans Ind Electron 59(6):2495–2502

17. Lee D-H, Pham TH, Ahn J-W (2013) Design and operation characteristics of four-two pole high-speed SRM for torque ripple reduction. IEEE Trans Ind Electron 60(9):3637–3643
18. Flieller D, Nguyen NK, Wira P, Sturtzer G, Abdeslam DO, Merckle J (2014) A self-learning solution for torque ripple reduction for nonsinusoidal permanent-magnet motor drives based on artificial neural networks. IEEE Trans Ind Electron 61(2):655–666
19. Lei G, Xu W, Hu JF, Zhu JG, Guo YG, Shao KR (2014) Multi-level design optimization of a FSPMM drive system by using sequential subspace optimization method. IEEE Trans Magn 50(2). Article 7016904
20. Yao D, Ionel DM (2013) A review of recent developments in electrical machine design optimization methods with a permanent-magnet synchronous motor benchmark study. IEEE Trans Ind Appl 49(3):1268–1275
21. Hasanien HM, Abd-Rabou AS, Sakr SM (2010) Design optimization of transverse flux linear motor for weight reduction and performance improvement using response surface methodology and genetic algorithms. IEEE Trans Energy Convers 25(3):598–605
22. Buja GS, Kazmierkowski MP (2004) Direct torque control of PWM inverter-fed AC motors—a survey. IEEE Trans Ind Electron 51(4):744–757
23. Kouro S, Cortes P, Vargas R, Ammann U, Rodriguez J (2009) Model predictive control—a simple and powerful method to control power converters. IEEE Trans Ind Electron 56 (6):1826–1838
24. Bolognani S, Bolognani S, Peretti L, Zigliotto M (2009) Design and implementation of model predictive control for electrical motor drives. IEEE Trans Ind Electron 56(6):1925–1936
25. Xia CL, Wang YF, Shi TN (2013) Implementation of finite-state model predictive control for commutation torque ripple minimization of permanent-magnet brushless DC motor. IEEE Trans Ind Electron 60(3):896–905
26. Morel F, Lin-Shi XF, Retif J-M, Allard B, Buttay C (2009) A comparative study of predictive current control schemes for a permanent-magnet synchronous machine drive. IEEE Trans Ind Electron 56(7):2715–2728
27. Liu C-H, Hsu Y-Y (2010) Design of a self-tuning PI controller for a STATCOM using particle swarm optimization. IEEE Trans Ind Electron 57(2):702–715
28. Zhu JG, Guo YG, Lin ZW, Li YJ, Huang YK (2011) Development of PM transverse flux motors with soft magnetic composite cores. IEEE Trans Magn 47(10):4376–4383
29. Guo YG, Zhu JG, Watterson PA, Wei Wu (2006) Development of a PM transverse flux motor with soft magnetic composite core. IEEE Trans Energy Convers 21(2):426–434
30. Wang TS, Zhu JG, Zhang YC (2011) Model predictive torque control for PMSM with duty ratio optimization. In: Proceedings of the 2011 International Conference on Electrical Machines and Systems (ICEMS), 20–23 Aug 2011, pp 1–5
31. Wang SH, Meng XJ, Guo NN, Li HB, Qiu J, Zhu JG et al (2009) Multi-level optimization for surface mounted PM machine incorporating with FEM. IEEE Trans Magn 45(10):4700–4703
32. Meng XJ, Wang SH, Qiu J, Zhang QH, Zhu JG, Guo YG, Liu DK (2011) Robust multi-level optimization of PMSM using design for six sigma. IEEE Trans Magn 47(10):3248–3251

Chapter 6
Design Optimization for High Quality Mass Production

Abstract In the last two chapters, the design optimization methods under the framework of deterministic approach were presented for electrical machines and drive systems. By the deterministic approach, all material and structural parameters in the manufacturing process are exact values that do not have any variations from their nominal values. However, there are many unavoidable uncertainties or variations in the industrial manufacturing process of electrical machines and drive systems, including mainly material diversity, manufacturing errors and assembly inaccuracy. These will result in big variations affecting the reliability and quality of electrical machines and drive systems in mass production. These variations are not investigated in the deterministic approach. The main aim of this chapter is to present a robust approach based on the technique of design for six-sigma (DFSS) for the design optimization of high-performance and high-quality electrical machines and drive systems in mass production. Meanwhile, two multi-level optimization strategies are presented to improve the optimization efficiency for high dimensional problems. Through the investigation of several design examples, it is shown that the reliability and quality of the investigated electrical machines and drive system can be increased greatly by using the proposed robust approach.

Keywords Design for six-sigma · Electrical drive systems · Model predictive control · Robust design optimization · System-level design optimization · Multi-objective optimization · Transverse flux machine · Permanent magnet motors · Monte carlo analysis

6.1 Introduction

Chapters 4 and 5 presented several design optimization methods for electrical machines in terms of different optimization situations, including multi-objective and high dimensional situations, and the system-level design optimization methods for electrical drive systems, including single- and multi-level optimization methods,

G. Lei et al., *Multidisciplinary Design Optimization Methods for Electrical Machines and Drive Systems*, Power Systems,
DOI 10.1007/978-3-662-49271-0_6

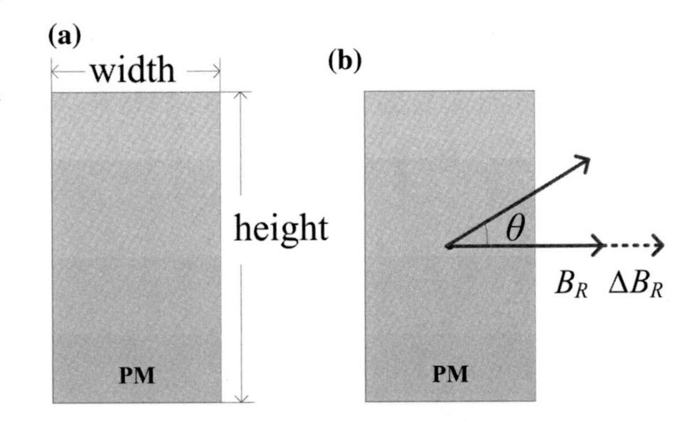

Fig. 6.1 Manufacturing tolerances for PMs, **a** dimension, **b** magnetization faults of magnitude B_R, and direction θ

respectively. Through the investigation of several design examples, it is shown that these proposed methods are efficient. For example, the multi-level optimization method can increase the steady-state and dynamic motor performances, such as higher output power and efficiency, and lower speed overshoot [1].

However, these proposed methods are a kind of deterministic design approach from the industrial design perspective and have not investigated the unavoidable variations (similar to the term of noise factors used in communication field) in the engineering manufacturing, including mainly material diversity, manufacturing error and assembly inaccuracy, and system parameter variations in practical operation environment [2, 3].

For example, the manufacturing quality of permanent magnets (PMs) is crucial to the performance of PM motors. There are at least two kinds of variations in the manufacturing of PMs. As shown in Fig. 6.1, the first one is the dimension, such as the height and width, and the second one is the magnetization faults of magnitude (B_R) and direction (θ) [4–6]. In [4], a practical example about the measurement data of PM width for a batch of 2000 PMs was presented. These PMs were from 3 manufacturing groups with the same lower limit (about 14.6 mm) and upper limit (about 14.7 mm). After the measurement, it was found that the average of one group (about 1000 PMs) is obviously smaller than the lower limit, and there is about 0.05 mm deviation from the average.

The problem mentioned above is really a challenge in both research and industrial communities as it includes not only the theoretical multi-disciplinary design and analysis (such as electromagnetic, thermal and mechanical analysis and power electronics), but also the practical engineering manufacturing of electrical machines and drive systems.

On the other hand, many new control algorithms have been developed for motor drives, e.g. model predictive control (MPC) and its improvements. Many algorithm parameters need to be optimized for good dynamic drive system performance. From the industrial application perspective, it is a natural requirement that the obtained optimal control algorithm parameters are robust against the variations of motor performance parameters due to the variations existing in material characteristics and manufacturing process. This issue is crucial for the batch production of novel drive systems [2, 7–9].

In order to find effective ways to deal with this problem, several robust design optimization methods have been investigated, such as Taguchi method [10–15] and six-sigma robust optimization method [16–19]. These two methods have been widely used to optimize the motor performance (including torque ripple, cost and output power) and quality against the manufacturing tolerances.

In our previous work, a six-sigma robust optimization method was first proposed to investigate a PM transverse flux machine (TFM) and a surface mounted PM synchronous machine (PMSM), respectively, by using different optimization algorithms [17–19]. Meanwhile, a drive system combining a PM TFM with a field oriented control (FOC) scheme was investigated in [3]. From the discussion, some interesting results have been obtained and it was that the system reliability has been improved significantly.

However, that work is only a case study. Only 6 parameters (4 motor structural parameters plus 2 control algorithm parameters) were investigated, and all of them are optimized at the same time. This method may be workable for some low dimensional problems, e.g. dimension D is smaller than 6, but it is very hard or ineffective for high dimensional problems due to the huge computational cost. Unfortunately, the practical drive systems are always high dimensional and D is often larger than 10.

The huge computational cost mentioned above is mainly from two parts. The first one is the finite element analysis (FEA) of motor and simulation of control algorithm required by the optimization algorithm. For example, considering a drive system optimization problem with 14 parameters, about 200*5*14 = 14,000 points are required if the differential evolution algorithm (DEA) is used as the optimization algorithm [3], where 5*14 is the population size and 200 is the iteration number of DEA. The second one is the sample size of Monte Carlo analysis (MCA) used to obtain the mean and standard deviation terms of objective and constraints in the robust model (see model (6.3) in the next section). Generally, this sample size is a large number, e.g. 10,000, which means for each design option in those 14,000 points, 10,000 extra points need to be calculated for MCA. Therefore, robust optimization for high dimensional problems can be a real challenge.

To solve the aforementioned questions, a systemic study was presented for this general and fundamental research topic. A design example of an electrical drive system has been investigated. This chapter also presents a robust approach for the design optimization of an electrical drive system based on the design for six-sigma (DFSS) technique. Section 6.2 describes the robust technique of DFSS. Section 6.3 presents two robust optimization methods for electrical machines in terms of single- and multi-objective situations. Section 6.4 presents the system-level robust design optimization methods for electrical drive systems, followed by the summary.

6.2 Design for Six-Sigma

In general, there are three kinds of design approaches from industry perspective, namely deterministic, reliability and robust approaches. Figure 6.2 illustrates a general optimization flowchart and features for them.

Figure 6.2a illustrates the optimization solutions obtained from the robust and deterministic approaches, respectively. As shown, the optimal solution obtained by deterministic approach (namely function minimum) is smaller than the robust optimum. However, when a variation or noise Δx happens, the objective's variation $\Delta f(x)$ of the deterministic approach is obviously larger than that of the robust approach. Most importantly, some variations from deterministic approach violate the basic constraints of design problem, e.g. lying inside the infeasible domain, and this is prohibited in engineering design [2].

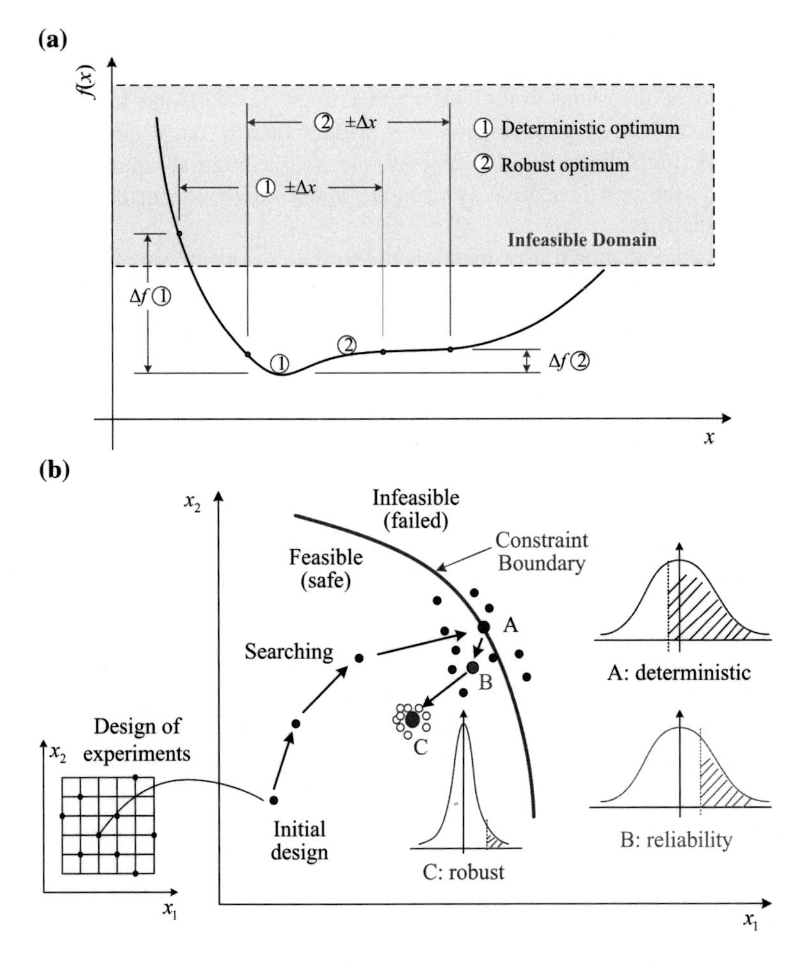

Fig. 6.2 Illustrations of deterministic, reliability and robust design approaches

Figure 6.2b illustrates the optimizing flowcharts for these three design approaches. As shown, the distance between constraint boundary and the solution obtained from robust approach is the furthest one, which means the reliability of the product is the highest. Meanwhile, the objective's variance of robust approach is the smallest, which means that the quality variance of robust approach is also the smallest and the products' quality is the highest. Generally, if a design scheme is not robust, it may be very difficult or even impossible to manufacture (e.g. requiring extreme material characteristics or unrealistically high manufacturing precision) due to current manufacturing technology or to operate (e.g. unstable system performance in the application environment).

As shown in Fig. 6.2b, the deterministic approach tends to push a design toward one or more constraint boundaries until those constraints are reached, which provides a high-risk design to the designer. Furthermore, the deterministic approach tends to search for the valley solutions or global minimal values from the point of view of mathematics. However, the valley point is highly sensitive to design parameter variations, i.e. the product performance will be degraded significantly in practical industry manufacturing [2]. Therefore, the robust approach is very important for modern quality control and design and should be taken into account in the system-level design optimization of drive systems.

Generally, a deterministic design with respect to an objective $f(\mathbf{x})$ and m constraints $g(\mathbf{x})$ has the form as

$$\begin{array}{ll} \min: & f(\mathbf{x}) \\ \text{s.t.} & g_i(\mathbf{x}) \leq 0,\ i = 1, \ldots, m, \\ & \mathbf{x}_l \leq \mathbf{x} \leq \mathbf{x}_u \end{array} \tag{6.1}$$

where $\mathbf{x}_l$ and $\mathbf{x}_u$ are the boundaries of design parameter $\mathbf{x}$ which is deterministic and does not cover any uncertain information. As mentioned above, there are many unavoidable noise factors in the industrial design and manufacturing process, such as the assembly tolerances and manufacturing imperfections in mass-production [4–6, 20–23]. Therefore, reliability design is developed to include the noise factors in the constraints to improve the reliability of products, in which $g(\mathbf{x})$ is converted to a probability function as

$$P_f = P(g(\mathbf{x}) > 0) \leq P^U, \tag{6.2}$$

where $\mathbf{x}$ is a vector of random variables, P_f the failure probability, and P^U its upper bound [2].

However, reliability design just focuses on the constraint boundary, and does not consider the variations of objectives and constraints in terms of those noise factors. Therefore, the quality distribution and average product performance cannot be evaluated. Fortunately, DFSS technique can deal with these problems very well. Actually, DFSS is a kind of robust design approach which was originated from the Six-Sigma Methodology developed by *MOTOROLA* and *GE* [16]. It is generally used to develop products to meet customer needs with very low defect levels. It has the design form as

$$\begin{aligned} \min: \quad & F[\mu_f(\mathbf{x}), \sigma_f(\mathbf{x})] \\ \text{s.t.}: \quad & g_i[\mu_f(\mathbf{x}), \sigma_f(\mathbf{x})] \leq 0, \; i = 1, \ldots, m \\ & \mathbf{x}_l + n\sigma_{\mathbf{x}} \leq \mu_{\mathbf{x}} \leq \mathbf{x}_u - n\sigma_{\mathbf{x}} \\ & \text{LSL} \leq \mu_f \pm n\sigma_f \leq \text{USL} \end{aligned}, \tag{6.3}$$

where μ and σ are the mean and standard deviation of the corresponding terms which are generally estimated by MCA method, LSL and USL the lower and supper specification limits, n is the sigma level, which is generally equivalent to a probability of a standard normal distribution as shown in Fig. 6.3.

Table 6.1 tabulates the equivalent percentage/probability for each sigma level. For example, 3σ means that the probability of $P(-3\sigma \leq x \leq 3\sigma)$ is 99.73 % assuming that x follows a standard normal distribution with mean 0 and variation 1 ($\sigma = 1$). In other words, 3σ is equivalent to a probability of 99.73 % or the POF is 0.27 %. This probability was deemed acceptable in statistical terms, and this value can be regarded as the quality control of short term, which means that there are 2,700 defects per million products.

However, with the development of long term quality control and management, this 3σ quality level is insufficient from the manufacturing perspective. From the rich experience of *MOTOROLA*, *GE* and others, an approximate 1.5 sigma shift in the mean (as shown in Fig. 6.4) was observed and this has been used to define the long term sigma quality as opposed to the above short term sigma quality [2, 16–18, 24–27].

For example, if there is a 1.5σ shift for 3σ quality control, the equivalent probability is

$$P(-4.5 \leq x \leq 1.5) = 93.3107\,\% \tag{6.4}$$

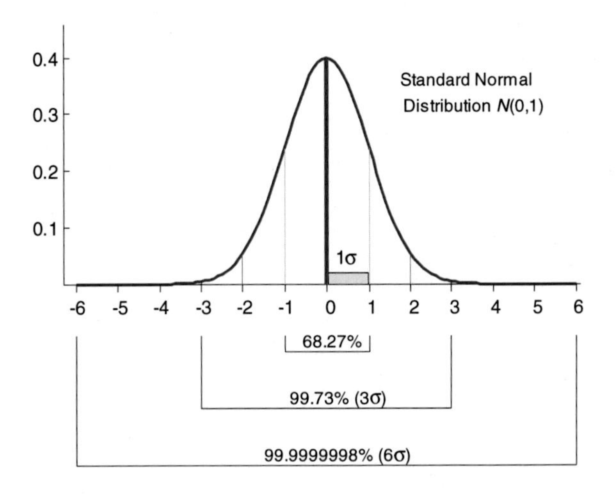

Fig. 6.3 Sigma level and its equivalent probability for a normal distribution

Table 6.1 Percentages and defects per million in terms of sigma level

Sigma level (σ)	Percentage	Defects per million (short term)	Defects per million (long term)
1	68.26	317,400	697,700
2	95.46	45,400	308,733
3	99.73	2,700	66,803
4	99.9937	63	6,200
5	99.999943	0.57	233
6	99.9999998	0.002	3.4

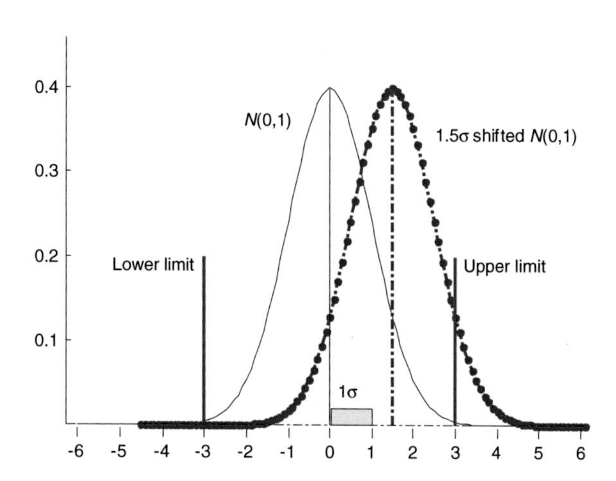

Fig. 6.4 Illustration of a 1.5σ shift in the mean for a normal distribution

Therefore, the failure rate is 6.6803 %, which means that the 3σ quality control is actually equivalent to 66,803 defects per million products being manufactured in the long term quality control technique. Obviously, this quality is not acceptable for the mass production of a product in industry. Similarly, 4σ and 5σ are equivalent to 6,200 and 233 defects per million, respectively, and they are not good choices for quality control either. To achieve the highest profit, 6σ level should be selected as it is equivalent to only 3.4 defects per million products, and it has been adopted in many companies worldwide nowadays.

To compare the product reliability by using different design approaches, a criterion called product probability of failure (POF) was used in many works. Assuming that all constraints in (6.3) are independent events, and then according to the Multiplication Theorem of Probability, the POF of the system described by (6.3) has the form as

$$\mathrm{POF} = 1 - \prod_{i=1}^{m} P(g_i \leq 0) \tag{6.5}$$

where $P(g_i \leq 0)$ means that the probability of event constraint g_i is correct for all samples in the MCA [2].

6.3 Robust Design Optimization of Electrical Machines

This section will investigate two optimization situations of single and multi-objectives, based on example study for the robust design of electrical machines.

6.3.1 *Single Objective Situation with a PM TFM*

6.3.1.1 Example on PM-SMC TFM

In this section, a PM TFM with SMC core (PM-SMC TFM) is investigated to illustrate the performance of the proposed method for electrical machines [28–30]. Figure 2.3 showed the prototype of this machine. It is designed to deliver a power of 640 W at 1800 rev/min. More details for this machine can be seen in Chaps. 2 and 4. From our design experience, eight structure parameters listed in Table 5.1 are significant to the performance of this machine [1]. They will be taken as the optimization parameters as well as variation parameters in the following robust optimization.

Another issue that has to investigate in the robust design of this PM-SMC TFM is the manufactuing quality of its SMC stator core. As the SMC core is compressed by a mould, the core density is related to the press size used. As mentioned in Sect. 4.8, the electromagnetic performance of this motor highly depends on this core density as shown in Fig. 4.33. Meanwhile, as shown in Fig. 4.34, the manufacturing cost of SMC cores directly depends on the selected press size. Consequently, the press size is a critical design variable as well as a variation factor for the evaluating of manufacturing quality of this machine.

Based on the above discussion, the optimization model for this machine can be defined as,

$$\begin{aligned} \min: \quad & f(\mathbf{x}) = Cost/C_0 + P_0/P_{out} \\ \text{s.t.} \quad & \begin{cases} g_1(\mathbf{x}) = 0.795 - \eta \leq 0, \\ g_2(\mathbf{x}) = 640 - \text{Pout} \leq 0, \\ g_3(\mathbf{x}) = sf - 0.8 \leq 0, \\ g_4(\mathbf{x}) = J_c - 6 \leq 0. \end{cases} \end{aligned} \tag{6.6}$$

where $\mathbf{x}$ is a vector of design parameters which include eight structure parameters and one manufacturing condition (indicated as x_9 in Table 6.2), C_0 and P_0 are the material cost and output power of the initial design scheme [1], η and P_{out} the motor efficiency and output power, sf and J_c the fill factor and current density of the winding, respectively.

With the robust optimization framework of (6.3), the robust optimization model of (6.6) can be expressed as

$$\begin{aligned} \text{min}: &\quad \mu_f(\mathbf{x}) \\ \text{s.t.}: &\quad \mu_{g_i}(\mathbf{x}) + 6\sigma_{g_i}(\mathbf{x}) \leq 0,\ i = 1,\ldots,4 \\ &\quad \mathbf{x}_l + 6\sigma_{\mathbf{x}} \leq \mu_{\mathbf{x}} \leq \mathbf{x}_u - 6\sigma_{\mathbf{x}} \\ &\quad \mu_f - 6\sigma_f \geq \text{LSL} \\ &\quad \mu_f + 6\sigma_f \leq \text{USL} \end{aligned} \tag{6.7}$$

MCA is used to estimate the mean and standard deviation terms in (6.7), and the sample size is 10^4. It should be noted that the optimization parameters in (6.6) and (6.7) are discrete values, and their step sizes are shown in Table 6.2 as well.

6.3.1.2 Optimization Results and Discussions

In the implementation, each parameter is defined to follow a normal distribution with standard deviation as 1/3 of its manufacturing tolerance. The tolerance values of the sixth and ninth motor parameters are defined as 1 % of their mean values. The tolerance values of other parameters are the same as their step sizes as shown in Table 6.2. To illustrate the performance of different methods, POF defined in (6.4) is taken as the criterion [17].

Tables 6.2 and 6.3 list the optimization results and the corresponding performance parameters obtained by two methods for this TFM, namely the deterministic design optimization of (6.6) and the robust design optimization of (6.7). Eight parameters shown in Table 5.1 are selected as the optimization parameters. Table 6.4 lists the robust levels for all constraints, and the POF values for the motor. Based on these results and comparison, the following conclusions can be drawn.

(1) For the deterministic design optimization, the obtained performance parameters of this machine include the cost of $27.8 and output power of 718 W. For the robust design optimization, the cost is $28.8 and output power 700 W. As shown, the robust design scheme has slightly higher cost and lower output power. Meanwhile, these values are better than those of the initial design scheme, which are $34.1 and 640 W, respectively.
(2) Considering the manufacturing cost, 200-ton press is suggested by the deterministic approach, the corresponding manufacturing cost of SMC core is $0.5/piece, and its density is 7.27 g/cm^3. However, 100-ton press is suggested by the robust approach, the manufacturing cost is only $0.2/piece, and density 6.60 g/cm^3. Therefore, lower manufacturing condition and cost are obtained by the robust method, and this is very important for the mass production.
(3) After MCA, the reliability of constraint g_4 is 50.37 % for the deterministic optimal scheme as shown in Table 6.4, and the corresponding sigma level is

Table 6.2 Robust optimization results for PM-SMC TFM

Par.	x_{m1}	x_{m2}	x_{m3}	x_{m4}	x_{m5}	x_{m6}	x_{m7}	x_{m8}	x_{m9}
Unit	Deg.	mm	mm	mm	mm	turn	mm	mm	ton
Step size	0.05	0.05	0.05	0.05	0.05	1	0.01	0.01	100
Deter.	10.95	7.35	8.00	7.00	9.05	115	1.30	0.95	200
Robust	10.00	8.30	8.15	7.30	9.90	118	1.30	0.97	100

Table 6.3 Performances for PM-SMC TFM after robust optimization

Par.	*cost*	η	P_{out}	*sf*	J_c	ρ
Unit	$	–	W	%	A/mm^2	g/cm^3
Deter.	27.8	0.82	718	60	6.00	7.27
Robust	28.8	0.83	700	59	5.76	6.60

Table 6.4 Sigma levels for constraints and POF for TFM after robust optimization

Par.	g_1	g_2	g_3	g_4	POF (%)
Deter.	6	6	6	50.37 %	49.63
Robust	6	6	6	6	~ 0

less than 1. Actually, the current density is 6.00 A/mm^2, which is the same as the limit of this constraint. As a result, the POF of motor is 49.63 %. For the robust scheme, the sigma levels for all constraints are larger than 6 and the POF is almost 0.

(4) Figures 6.5, 6.6, 6.7, 6.8 show the distributions of cost, output power, efficiency and current density respectively for both methods. As shown, the standard deviations of cost and output power of robust optimal scheme are smaller than those of the deterministic optimization scheme. As shown in Fig. 6.7, the robust optimal scheme can produce a larger mean and a smaller standard deviation for the efficiency of this TFM compared with the deterministic scheme.

As shown in Fig. 6.8, all current distribution points of robust design scheme are satisfied with the condition g_4 of "no larger than 6.0 A/mm^2. It can also be seen that many points of deterministic design scheme are not satisfied with this condition. Therefore, the reliability and sigma level of this constraint of deterministic method is very low, and the POF of motor is high. Actually, the lower cost of deterministic optimization is obtained at the cost of low reliability and robustness. This is not acceptable in engineering design.

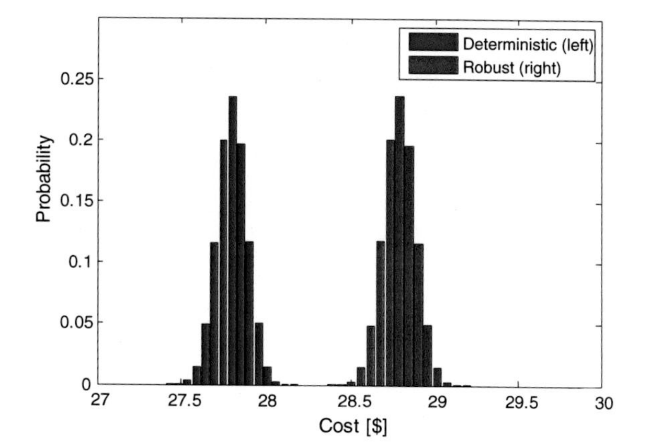

Fig. 6.5 Cost distributions of deterministic and robust optimal schemes

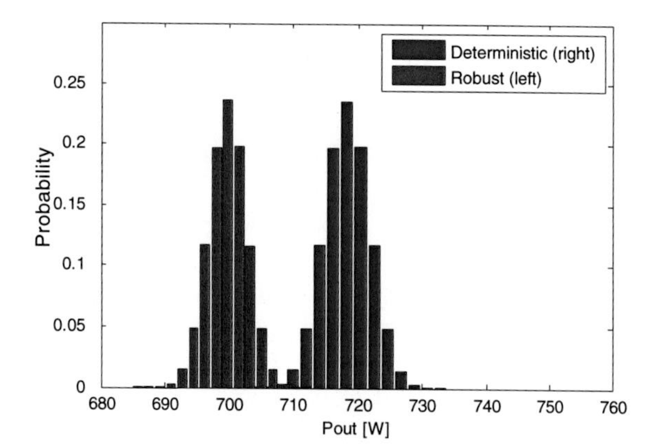

Fig. 6.6 Output power distributions of deterministic and robust optimal schemes

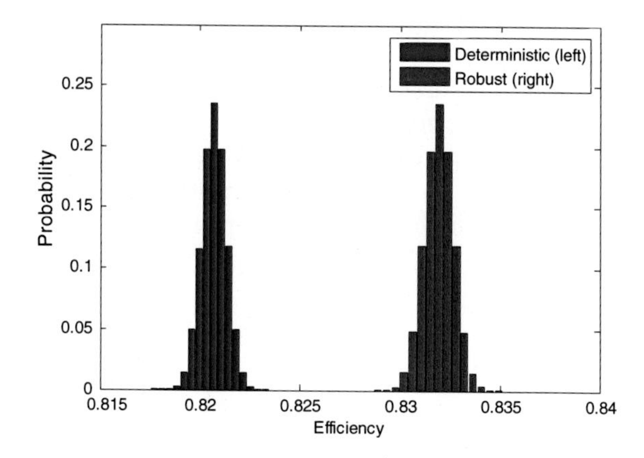

Fig. 6.7 Efficiency distributions of deterministic and robust optimal schemes

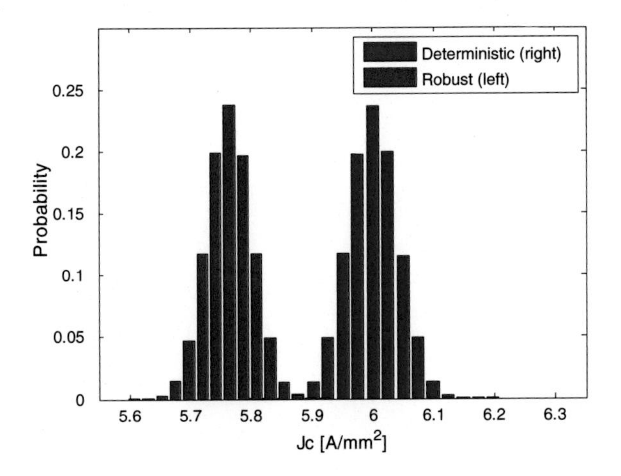

Fig. 6.8 Current density distributions of deterministic and robust optimal schemes

6.3.2 *Multi-objective Optimization with a PM TFM*

6.3.2.1 Multi-objective Robust Optimization Model

Section 6.3.1 presents a robust design optimization for a PM-SMC TFM under single-objective optimization situation. From the discussion, it can be found that the manufacturing quality of the motor has been increased significantly. However, two other issues should be investigated for the industrial applications of PM-SMC motors besides the robust analysis. Firstly, multi-objective design schemes are necessary as it is hard to determine the weights for different objectives without detailed information of industrial applications. Secondly, high computational cost is also an important issue as this is a high dimensional optimization problem and 3D finite element model (FEM) is involved. Therefore, this section presents an improved multi-objective sequential optimization method (MSOM) for the robust multi-objective design optimization of these PM-SMC motors to improve their industrial applications. The PM-SMC TFM discussed in the last section will be investigated here to illustrate the proposed new method.

Considering the manufacturing condition and material characteristic of SMC cores, the multi-objective design optimization model of PM-SMC motors can be defined as

$$\begin{array}{ll} \min: & \begin{cases} f_1(\mathbf{x}) = Cost \\ f_2(\mathbf{x}) = -P_{out} \end{cases} \\ \text{s.t.} & \begin{cases} g_1(\mathbf{x}) = 0.795 - \eta \leq 0, \\ g_2(\mathbf{x}) = 640 - P_{\text{out}} \leq 0, \\ g_3(\mathbf{x}) = sf - 0.8 \leq 0, \\ g_4(\mathbf{x}) = J_c - 6 \leq 0. \end{cases} \end{array}, \tag{6.8}$$

where $\mathbf{x}_s$, $\mathbf{x}_{mt}$, and $\mathbf{x}_{mf}$ are the structure, material and manufacturing parameters, respectively. To achieve the six sigma quality manufacturing, the design model can be converted into (6.9) within the framework of DFSS [2, 18].

$$\begin{array}{ll} \min: & \{F_k = \mu_{f_k}(\mathbf{x}),\ k = 1,2\} \\ \text{s.t.} & \mu_{g_i}(\mathbf{x}) + n\sigma_{g_i}(\mathbf{x}) \le 0,\ i = 1,\ldots,4 \end{array} \quad (6.9)$$

where μ and σ stand for the mean and standard deviation of the corresponding terms.

6.3.2.2 Improved Multi-objective Sequential Optimization Method

An MSOM has been presented for the multi-objective optimization of electrical machines in Sect. 4.4. However, it is hard to handle high dimensional problems for that MOSM. To solve this problem, this section presents an improved MSOM, which includes the following four steps:

(1) Generate an initial sample set S(0) and obtain an initial Pareto optimal solution P(0) by using the non-dominated sorting genetic algorithm (NSGA) II [31], a classical multi-objective optimization algorithm, which can be used to optimize models (6.8), (6.9) and their approximate models. To increase the optimization efficiency, the Kriging model will be employed to construct the optimization models of (6.8) and (6.9) in this work.
(2) Update the samples based on the obtained Pareto optimal solutions. Firstly, find the significant parameters by using the sensitivity analysis techniques introduced in Sect. 4.5, such as the local sensitivity analysis and design of experiments (DOE) techniques. Secondly, generate new samples by using a 2-level sampling method. Finally, update all the Kriging models.
(3) Optimize the obtained Kriging model [32–34] by using NSGA II, and get the updated Pareto optimal solution P(k).
(4) Compute the root mean square error (RMSE) of the obtained Pareto points for each Kriging model. If all RMSEs are no more than ε, output the solution, otherwise go to the second step.

6.3.2.3 Optimization Results and Discussion

Figures 6.9, 6.10, 6.11, 6.12 illustrate the optimization results for both deterministic and robust design approaches. The following conclusions can be drawn:

(1) Figure 6.9 illustrates the Pareto optimal solution obtained from deterministic model (6.8) and robust model (6.9), respectively. It can be found that the output power increases with the increase of cost and vice versa. The front of

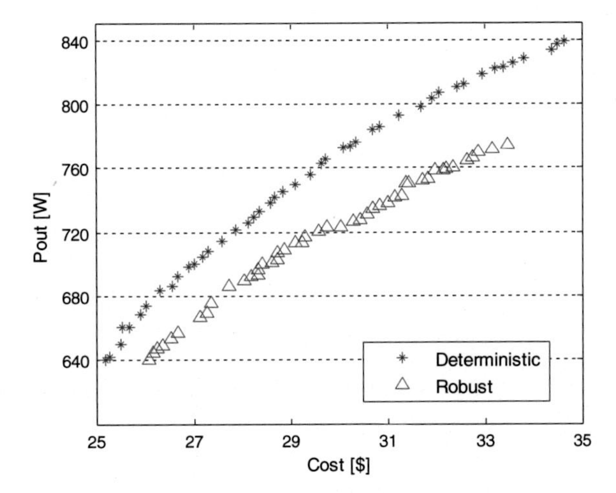

Fig. 6.9 Pareto optimal solutions for both deterministic and robust design approaches

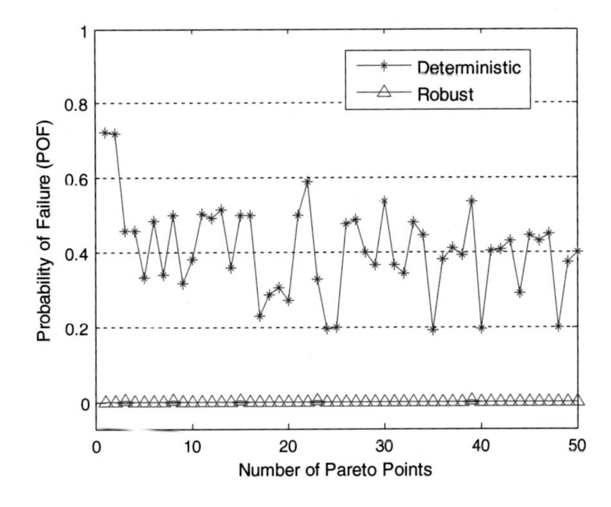

Fig. 6.10 POFs of the Pareto points of deterministic and robust design approaches

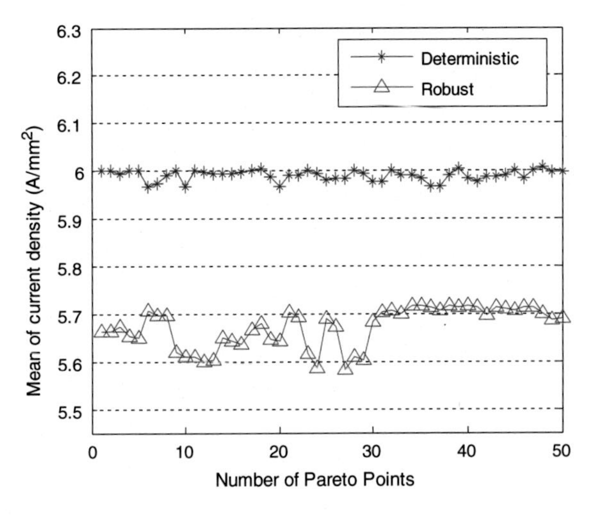

Fig. 6.11 Mean of current density of the Pareto points

Fig. 6.12 SMC core density of the Pareto points

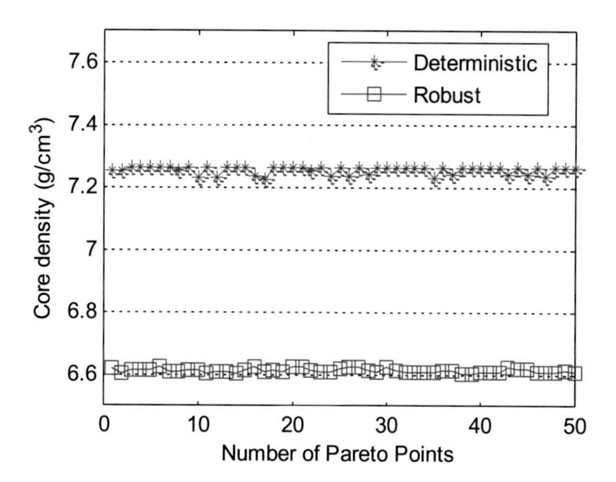

Pareto solutions obtained by the robust approach is lower than that by the deterministic approach, meaning that to achieve the same output power, the needed cost of robust design approach is higher than that of deterministic one.

(2) Figure 6.10 illustrates the POF values of all Pareto points for both approaches. It can be found that the POF values of deterministic design approach (or Pareto points) are unstable and higher than those of robust approach. Some of them are even over 50 % higher. These are bad designs from the point of view of high quality industrial design.

For the robust multi-objective design approach, the POF values are almost 0. Therefore, although the needed cost for the same output power of the deterministic scheme is less than that of the robust approach, its lower cost is achieved at the cost of lower POF.

(3) Figure 6.11 shows the means of current density (J_c) for all Pareto points. It can be seen that deterministic designs have higher means of J_c. The means of J_c of the robust designs are obviously smaller than the limit of 6 A/mm^2, and the average of these means is 5.67 A/mm^2. However, many of the deterministic designs are beyond the limit, and the average is 5.99 A/mm^2. Thus, the POF values of g_4 of the deterministic approach are higher than those of the robust approach. For other constraints, the POFs and means for all Pareto points are also obtained, but not much difference has been founded. Therefore, the current density issue is the main reason why the deterministic approach has higher POFs than the robust approach as shown in Fig. 6.10.

(4) Figure 6.12 illustrates the core density for all Pareto points. It can be found that the core densities of deterministic designs are around 7.2 g/cm^3, which means that a 200-ton press is needed to compact these cores. On the other hand, the core densities of all robust designs are around 6.6 g/cm^3, and only a 100-ton press is required. Therefore, the robust approach needs lower manufacturing condition and cost than the deterministic approach.

(5) In terms of the computational cost, the direct optimization method (NSGA II with FEM) requires about 12,000 FEM points, in which half points were sampled for the no-load analysis and others were used for the operation analysis of this machine. The proposed method requires only about 3,800 FEM points, which is much less than that required by the direct optimization method.

In summary, the proposed robust multi-objective optimization method can significantly improve the reliability and manufacturing quality of the motor with lower manufacturing condition and cost. Consequentially, it will promote significantly the industrial applications of PM-SMC motors.

6.4 Robust Design Optimization of Electrical Drive Systems

6.4.1 Single-Level Robust Optimization Method

Figure 6.13 illustrates a block diagram of the system-level robust optimization method for electrical drive systems. Since all parameters are directly optimized at the system level, it is called the single-level robust optimization method.

It consists of the following three steps:

Step 1: Determination of system-level robust optimization model

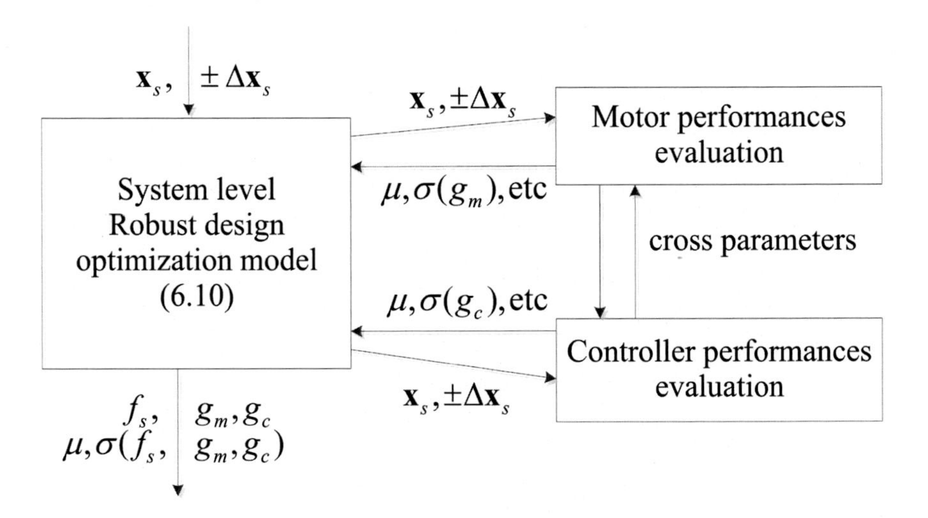

Fig. 6.13 Single level robust optimization method for drive systems

Based on the deterministic form of system-level optimization model (5.3), the system-level robust optimization model of drive system can be expressed as

$$\begin{array}{ll} \min: & F[\mu_{f_s}(\mathbf{x}_s),\ \sigma_{f_s}(\mathbf{x}_s)] \\ \text{s.t.}: & \mu_{g_{mi}}(\mathbf{x}_s) + n\sigma_{g_{mi}}(\mathbf{x}_s) \le 0,\ i = 1, \ldots, N_m \\ & \mu_{g_{cj}}(\mathbf{x}_s) + n\sigma_{g_{cj}}(\mathbf{x}_s) \le 0,\ j = 1, \ldots, N_c \\ & \mathbf{x}_{sl} + n\sigma_{\mathbf{x}_s} \le \mu_{\mathbf{x}_s} \le \mathbf{x}_{su} - n\sigma_{\mathbf{x}_s} \\ & \mathrm{LSL} \le \mu_{f_s} \pm n\sigma_{f_s} \le \mathrm{USL} \end{array} . \tag{6.10}$$

To estimate the mean and standard deviation terms in (6.10), each design parameter x_i in $\mathbf{x}_s$ is assumed normally distributed as $N(x_i, \sigma_{x_i}^2)$ with $\sigma_{x_i} = \Delta x_i/3$, where Δx_i is the manufacture tolerance of x_i.

Step 2: Selection of an optimization method for model (6.10)

As drive systems are always high dimensional and non-linear design problems, intelligent algorithms can be good choices in many situations, such as the genetic algorithm (GA) and DEA [1–3, 35–37].

Step 3: Implementation of optimization process

Firstly, determine the manufacture tolerance for $\mathbf{x}_s$ and obtain the distribution parameters for each parameter. Secondly, generate an initial population of $\mathbf{x}_s$ and its noise population. Thirdly, evaluate the steady-state and dynamic performance parameters of the drive system. Meanwhile, the objectives, constraints in (6.10), and their means and standard deviations can be gained by using the MCA method. Finally, apply the optimization algorithm until the convergence criterion is met.

Because, as mentioned above, the computational cost of this single-level method is always huge as these design problems are generally high dimensional and non-linear, the computational cost of whole system optimization is very expensive, in which the major part is the computational cost of FEM for the motors. To solve this problem, a multi-level robust optimization method is proposed.

6.4.2 *Multi-level Robust Optimization Method*

6.4.2.1 Method Description

Figure 6.14 illustrates the framework of multi-level robust optimization method for drive systems, which consists of three levels, namely the motor, control, and system levels. In the implementation, the first step is to define the deterministic and robust optimization models for the motor and control levels, respectively. For the motor level, its deterministic optimization model has been defined in Chap. 5, and has the form as

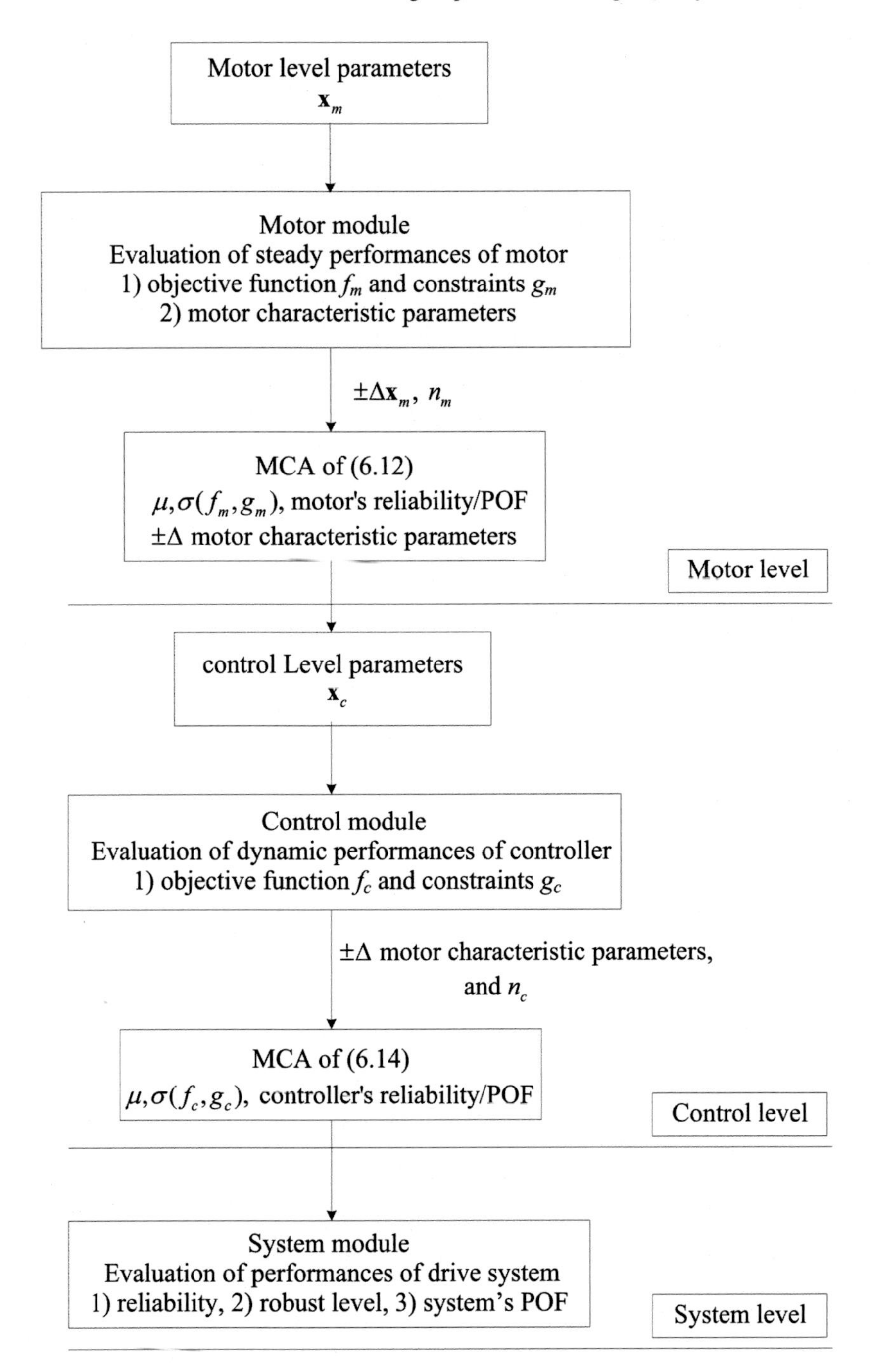

Fig. 6.14 Multi-level robust optimization framework for drive systems

$$\begin{aligned} \min: &\quad f_m(\mathbf{x}_m) \\ \text{s.t.} &\quad g_{mi}(\mathbf{x}_m) \le 0,\ i = 1, \ldots, N_m\,, \\ &\quad \mathbf{x}_{ml} \le \mathbf{x}_m \le \mathbf{x}_{mu} \end{aligned} \tag{6.11}$$

where $\mathbf{x}_m$, f_m and g_m are the design parameter vector, objective and constraint, and $\mathbf{x}_{ml}$ and $\mathbf{x}_{mu}$ the lower and upper boundaries of $\mathbf{x}_m$, respectively.

With the DFSS technique, the corresponding robust model can be expressed as

$$\begin{aligned} \min: &\quad F[\mu_{f_m}(\mathbf{x}_m),\ \sigma_{f_m}(\mathbf{x}_m)] \\ \text{s.t.}: &\quad \mu_{g_{mi}}(\mathbf{x}_s) + n_m\sigma_{g_{mi}}(\mathbf{x}_s) \le 0,\ i = 1, \ldots, N_m \\ &\quad \mathbf{x}_{ml} + n_m\sigma_{\mathbf{x}_m} \le \mu_{\mathbf{x}_m} \le \mathbf{x}_{mu} - n_m\sigma_{\mathbf{x}_m} \\ &\quad \text{LSL} \le \mu_{f_m} \pm n_m\sigma_{f_m} \le \text{USL} \end{aligned}, \tag{6.12}$$

where n_m is the sigma level specified for the motor level. It should be noted that the required sigma levels for motor level and control level may be different from that in system level model, and different symbols are specified for them.

For the control level, its deterministic design optimization model has the form as

$$\begin{aligned} \min: &\quad f_c(\mathbf{x}_c) \\ \text{s.t.} &\quad g_{cj}(\mathbf{x}_c) \le 0,\ j = 1, \ldots, N_c\,, \\ &\quad \mathbf{x}_{cl} \le \mathbf{x}_c \le \mathbf{x}_{cu} \end{aligned} \tag{6.13}$$

where $\mathbf{x}_c$, f_c and g_c are the design parameter vector, objective and constraint, and $\mathbf{x}_{cl}$ and $\mathbf{x}_{cu}$ the lower and upper boundaries, respectively.

Similarly, its robust model can be expressed as

$$\begin{aligned} \min: &\quad F[\mu_{f_c}(\mathbf{x}_c),\ \sigma_{f_c}(\mathbf{x}_c)] \\ \text{s.t.}: &\quad \mu_{g_{cj}}(\mathbf{x}_c) + n_c\sigma_{g_{cj}}(\mathbf{x}_c) \le 0,\ j = 1, \ldots, N_c \\ &\quad \mathbf{x}_{cl} + n_c\sigma_{\mathbf{x}_c} \le \mu_{\mathbf{x}_c} \le \mathbf{x}_{cu} - n_c\sigma_{\mathbf{x}_c} \\ &\quad \text{LSL} \le \mu_{f_c} \pm n_c\sigma_{f_c} \le \text{USL} \end{aligned}, \tag{6.14}$$

where n_c is the sigma level specified for the control level.

Based on the above robust optimization models for motor, control, and system levels, the proposed multi-level robust optimization method shown in Fig. 6.14 can be implemented in the following four steps:

Step 1: Determination of the POF values for motor, control and system levels, respectively, in terms of design requirements and available manufacturing conditions
Step 2: Optimization of motor level with model (6.12)

Besides the motor performance parameters, such as output power and efficiency, some characteristic parameters of the motor, such as winding resistance and inductance and PM flux, should be calculated at the same time, because they will be considered as the noise factors in the optimization of control level.

Step 3: Optimization of control level with model (6.14)

The main aim of this step is to evaluate the fluctuations of dynamic performance with respect to the noise factors from the motor characteristic parameters.

Step 4: Performance evaluation of system level

Here are two remarks for this multi-level robust optimization method. Firstly, if the dimension of motor level or control level is large, the optimization of corresponding level can be divided into two or three sub-levels. Secondly, as mentioned before, the optimization process is usually quite time-consuming because of the huge computational costs of FEM for electromagnetic analysis of the motor and the MCA process in robust optimization. Approximate models, such as the response surface model and Kriging model, can be used to replace the FEM [32–34]. The Kriging model will be employed for the design example in this work.

6.4.2.2 Design Example of a Drive System with TFM and MPC

A. Optimization model for motor level

The optimization objective is to minimize the material cost while still keeping or improving its performance compared with the initial design. The optimization model is

$$\begin{aligned} \min:\ & f_m(\mathbf{x}_m) = w_1 \frac{Cost}{C_0} + w_2 \frac{P_0}{P_{\mathrm{out}}} \\ \text{s.t.}:\ & g_{m1}(\mathbf{x}_m) = 0.795 - \eta \le 0, \\ & g_{m2}(\mathbf{x}_m) = 640 - P_{\mathrm{out}} \le 0, \\ & g_{m3}(\mathbf{x}_m) = sf - 0.8 \le 0, \\ & g_{m4}(\mathbf{x}_m) = J_c - 6 \le 0, \\ & \mathbf{x}_{ml} \le \mathbf{x}_m \le \mathbf{x}_{mu} \end{aligned} \tag{6.15}$$

where C_0 and P_0 are the cost and output power (P_{out}) of the initial prototype, η is motor efficiency, sf the winding fill factor, and J_c the current density of copper wire winding. The optimization parameters are illustrated in Table 5.1 and Fig. 5.5.

B. Optimization model for control level

Figure 2.37 showed a diagram of this improved MPC control scheme. There are two important modules needed to be designed and optimized in this improved MPC, namely cost function and duty ratio module [38–40]. Six parameters presented in Sect. 5.4.2 should be optimized in the control level. They are A, N, C_T, C_ψ,

K_p and K_i, where K_p and K_i are the parameters of PI controller. One objective and four constraints are considered for this level as the following

$$\begin{array}{ll} \min: & f_c(\mathbf{x}_c) = w_3 \frac{\mathrm{RMSE}(T)}{T_{rated}} + w_4 \frac{\mathrm{RMSE}(n)}{n_{rated}} + w_5 n_{os} \\ \text{s.t.}: & g_{c1}(\mathbf{x}_c) = \mathrm{RMSE}(T)/T_{rated} - 0.06 \le 0, \\ & g_{c2}(\mathbf{x}_c) = \mathrm{RMSE}(n)/n_{rated} - 0.05 \le 0, \\ & g_{c3}(\mathbf{x}_c) = n_{os} - 0.02 \le 0, \\ & g_{c4}(\mathbf{x}_c) = t_s - 0.02 \le 0, \\ & \mathbf{x}_{cl} \le \mathbf{x}_c \le \mathbf{x}_{cu} \end{array}, \tag{6.16}$$

where RMSE is the root mean square error of an item in the steady operation period, ω the motor speed, n_{os} the overshoot of speed, and t_s the settling time.

C. Robust optimization models

Firstly, based on (6.15), the robust optimization model for the motor level has the form as

$$\begin{array}{ll} \min: & \mu_{f_m}(\mathbf{x}_m) \\ \text{s.t.} & \begin{cases} \mu_{g_{mi}}(\mathbf{x}_m) + n_m \sigma_{g_{mi}}(\mathbf{x}_m) \le 0, \ i = 1, \ldots, 4 \\ x_{mlj} + n_m \sigma_{x_{mj}} \le \mu_{x_{mj}} \le x_{muj} - n_m \sigma_{x_{mj}}, \\ \qquad\qquad j = 1, \ldots, 8 \end{cases} \end{array}. \tag{6.17}$$

In the implementation, the manufacturing tolerances of motor parameters (x_{m1} to x_{m8}) are specified as 0.05 deg., 0.05 mm, 0.05 mm, 0.05 mm, 0.05 mm, 0.5 turn, 0.01*x_{m7} mm and 0.01 mm, respectively, according to the previous engineering experience.

Secondly, based on (6.16), the robust optimization model of control level can be defined as

$$\begin{array}{ll} \min: & \mu_{f_c}(\mathbf{x}_c) \\ \text{s.t.} & \begin{cases} \mu_{g_{ci}}(\mathbf{x}_c) + n_c \sigma_{g_{ci}}(\mathbf{x}_c) \le 0, \ i = 1, \ldots, 4 \\ x_{clj} \le x_{cj} \le x_{cuj}, \ j = 1, \ldots, 6 \end{cases} \end{array}. \tag{6.18}$$

It should be noted that the parameters of control level (x_{c1} to x_{c6}) are digital parameters and do not have disturbances for the MCA. There are only four robust constraints in (6.18).

However, there are four random variables in this model, namely the resistance (R), inductance (L), torque (T), and PM-flux corresponding to the random variables of all motor parameters x_{m1} to x_{m8}. After the MCA, the output parameters are the means and deviations of objectives and constraints of control level. Then, model (6.18) can be calculated and the POF of control level can be obtained.

D. Optimization flowchart, results and discussion

Figure 6.15 shows the multi-level robust optimization flowchart for this drive system. It mainly includes the following five steps:

Step 1: Specifying the expected sigma level or POF for the drive system

In this work, 6σ manufacturing quality is the expected quality level. Therefore, n_m and n_c are defined as 6.

Step 2: Dividing the initial design space into three subspaces/levels

Empirically, the eight parameters in the motor level can be divided into two levels. The first one (X1) includes PM parameters (x_{m1} and x_{m2}) and coil parameters (x_{m6} and x_{m7}), which are significant to the optimization objective. The others are non-significant factors and will be placed in the second level X2. Therefore, the optimization flowchart of the motor level is defined as two sublevels. In total, a three-level optimization framework can be obtained for the whole drive system with control as the third level.

Step 3: Optimizing the motor level

Firstly, let the parameters in X2 be the initial design dimensions and optimize the parameters in X1. Secondly, optimize the parameters in X2 with the obtained optimal parameters in X1. If the relative error of the objective is not smaller than a given value ε (the default is 1 %), update X2 and conduct the optimization again until $\Delta\mu(f_m)/\mu(f_m) \leq \varepsilon$ is met. After the optimization of motor level, the motor characteristic parameters, such as R, L and PM-flux, can be obtained as well, and they will be used as the input parameters in the control level.

Step 4: Optimizing the control level

The input parameters of this level are the algorithm parameters (x_{c1} to x_{c6}) in MPC and characteristic parameters obtained from the above motor level optimization. These characteristic parameters are also taken as the noise parameters in the optimization of this level.

Step 5: Calculating the POF for the whole system and outputting the optimization results.

Table 6.5 lists the optimization results for the TFM (motor level) obtained from three approaches, namely the initial, deterministic, and robust design approaches. Table 6.6 tabulates the optimization results for the improved MPC (control level) obtained by the deterministic and robust design approaches. Table 6.7 lists the reliabilities (column p) and robust levels (column σ) for all constraints, and the POF values for the motor, controller, and drive system, respectively. Based on these results and a comparison of them, the following conclusions can be drawn:

(1) By the deterministic approach based on models (6.15) and (6.16), the optimum motor efficiency, output power, and material cost are 81.3 %, 670 W, and \$26.9, respectively. After the MCA, the reliability of constraint g_{m4} is

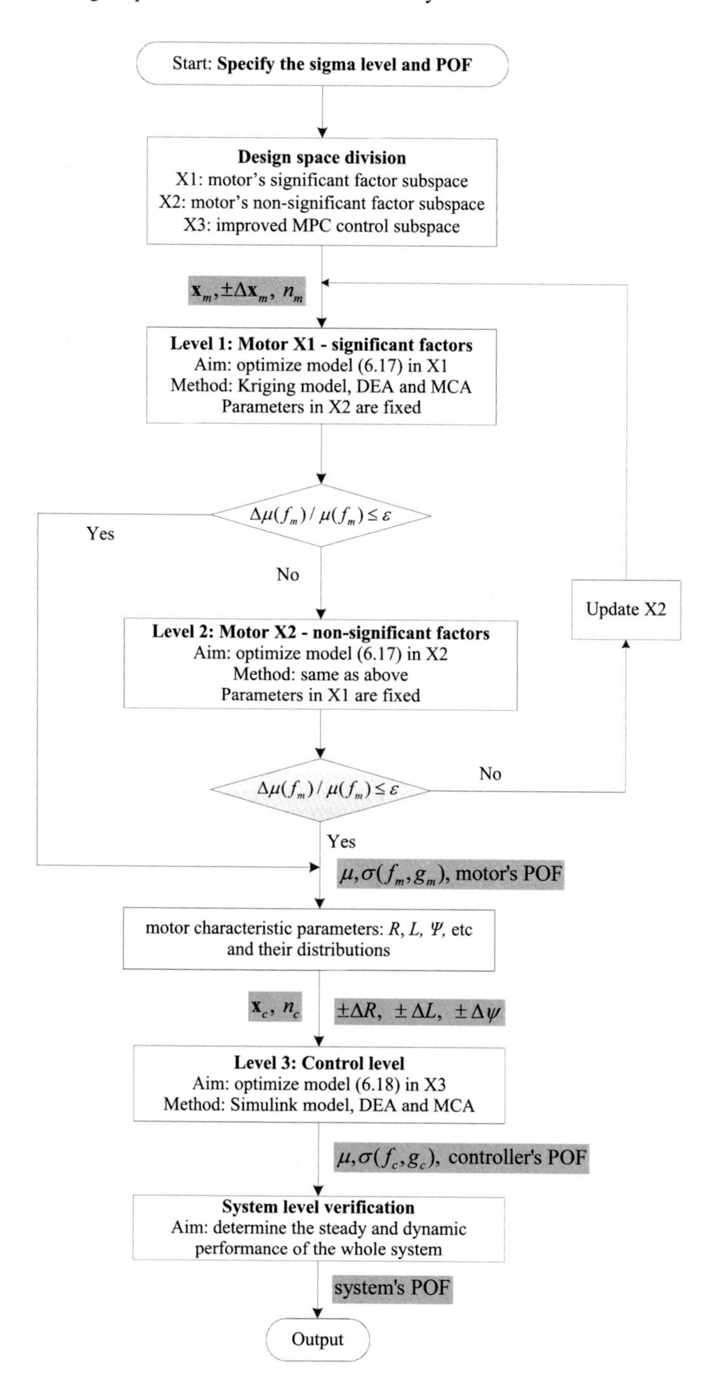

Fig. 6.15 Flowchart for the multi-level robust optimization method

Table 6.5 Optimization results of the TFM (motor level)

Par.	Nota.	Unit	Initial	Robust	Deterministic
x_{m1}	θ_{PM}	deg.	12	10.20	10.00
x_{m2}	W_{PM}	mm	9	7.90	7.65
x_{m3}	W_{stc}	mm	9	8.15	8.0
x_{m4}	W_{sta}	mm	8	8.05	7.95
x_{m5}	H_{str}	mm	10.5	10.85	10.9
x_{m6}	N_c	turn	125	111	110
x_{m7}	D_c	mm	1.25	1.27	1.27
x_{m8}	l_g	mm	1.0	0.97	0.95
η		–	79.5 %	81.1 %	81.3 %
P_{out}		W	640	671	670
sf		–	0.56	0.51	0.50
J_c		A/mm^2	4.72	5.75	5.96
$Cost$		\$	35.8	28.0	26.9
f_m		\$	–	1.68	1.65

Table 6.6 Optimization results of the improved MPC (control level)

Par.	Nota.	Robust	Deterministic
x_{c1}	A	0.806	0.386
x_{c2}	N	5	7
x_{c3}	C_T	0.882	1.17
x_{c4}	C_ψ	0.0366	0.03568
x_{c5}	K_i	0.116	0.23
x_{c6}	K_p	0.725	1.619
RMSE(T)/T_{rated}		4.64 %	3.95 %
RMSE(n)/n_{rated}		0.13 %	0.02 %
n_{os}		1.78 %	0.90 %
t_s		0.01	0.01
f_c		6.55 %	4.87 %

Table 6.7 Reliability and robust level of the drive system

Par.	Robust		Deterministic	
	p	σ	p	σ
g_{m1}	1	>6	1	>6
g_{m2}	1	>6	1	>6
g_{m3}	1	>6	1	>6
g_{m4}	1	>6	83.45 %	1.4
g_{c1}	1	>6	1	>6
g_{c2}	1	>6	99.90 %	3.3
g_{c3}	1	>6	1	>6
g_{c4}	1	>6	19.40 %	0.2
POF motor	~0.0 %		16.55 %	
POF control	~0.0 %		80.62 %	
POF system	~0.0 %		83.83 %	

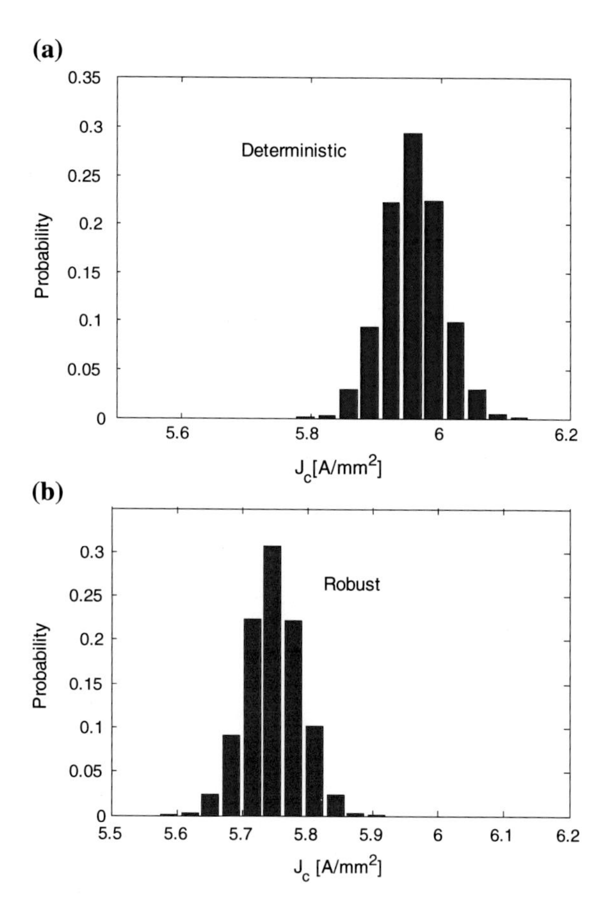

Fig. 6.16 Distributions of current density in winding, **a** deterministic approach, and **b** robust approach

83.45 %, and the corresponding sigma level is 1.4. As a result, the POF of motor is 16.55 %. As an example, Fig. 6.16 illustrates the current density distribution of copper wire corresponding to constraint g_{m4}. As shown, many points in Fig. 6.16a are beyond the limit of current density, 6 A/mm^2.

Meanwhile, the reliabilities of constraints g_{c2} and g_{c4} are 99.90 % and 19.40 %, and the corresponding sigma levels are 3.3 and 0.2, respectively. As a result, the POF of control level is 80.62 %. As another example, Fig. 6.17 illustrates the distribution of the settling time corresponding to constraint g_{c4}. As shown, most of the points violate constraint g_{c4} of "no more than 0.02 s after the load is applied". This results in a high POF for the control level. It should be noted that the last column (t = 0.14 s) in deterministic figure shows the probability for the issue that settling times are no less than 0.14 s instead of exactly equaling 0.14 s. Figure 6.18 illustrates the distribution of g_2, which is related to speed. As shown, there are also several points violating the limit "no more than 0.05".

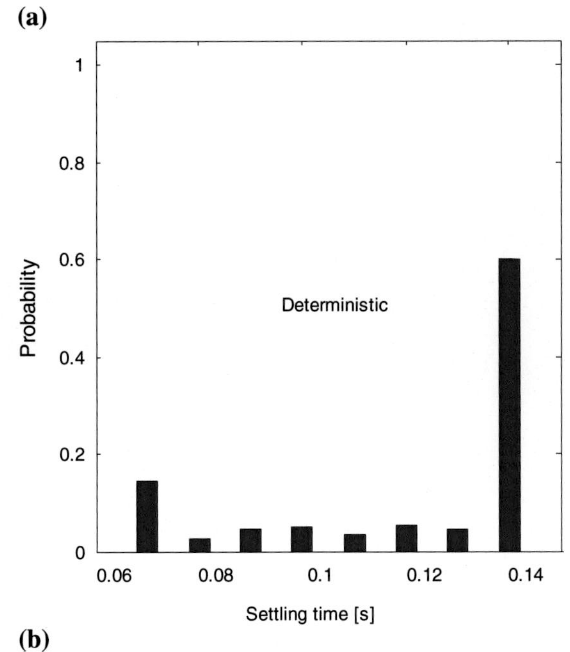

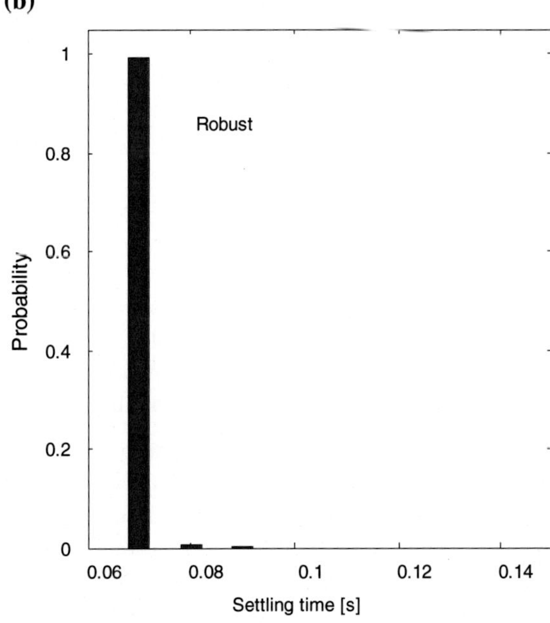

Fig. 6.17 Distributions of settling time (load is applied at time 0.07 s), **a** deterministic approach, and **b** robust approach

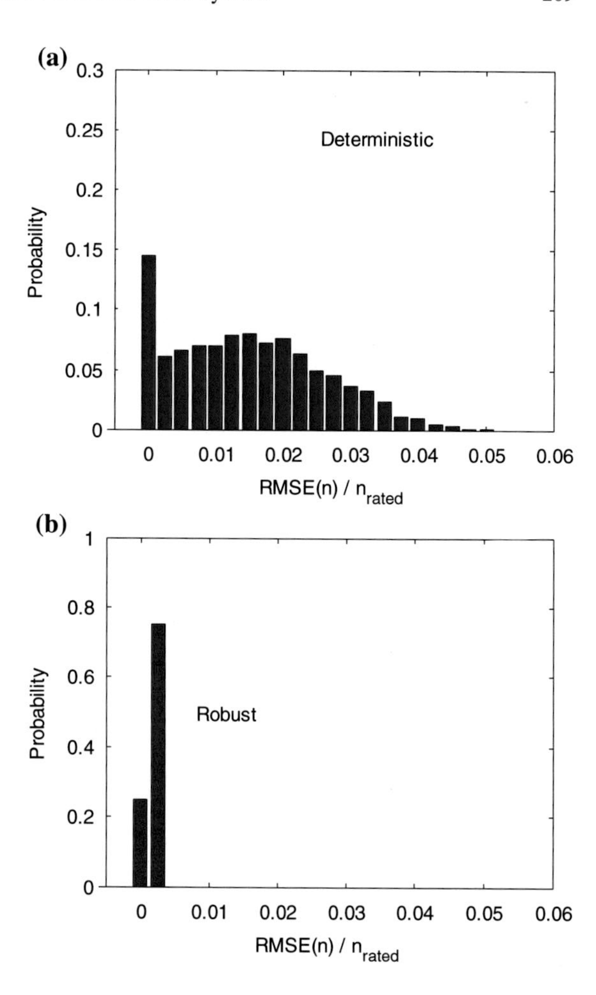

Fig. 6.18 Distributions of constraint g_2 related to speed, **a** deterministic approach, and **b** robust approach

Finally, the POF of the whole drive system is 83.83 %. This is absolutely not an acceptable system design for engineering applications.

(2) By the robust approach, three iteration processes are required to get the optimal results of the multi-level optimization method. Figure 6.19 shows the iteration process of multi-level optimization for the motor level. As shown, level 1 is optimized twice while level 2 is optimized only once. The optimum motor efficiency is 81.1 %, and output power 671 W, which are close to those obtained by the deterministic approach, whereas the material cost is \$28, which is bigger than that of deterministic design. Regarding the control level, as shown in Table 6.6, the dynamic performance of this drive system after robust optimization is slightly worse than that of the deterministic optimization. However, all comply well with those constrains in control level.

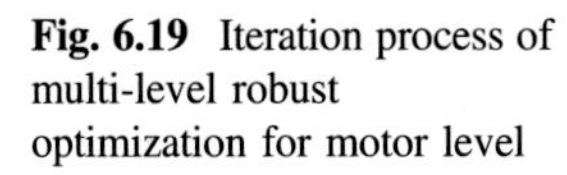

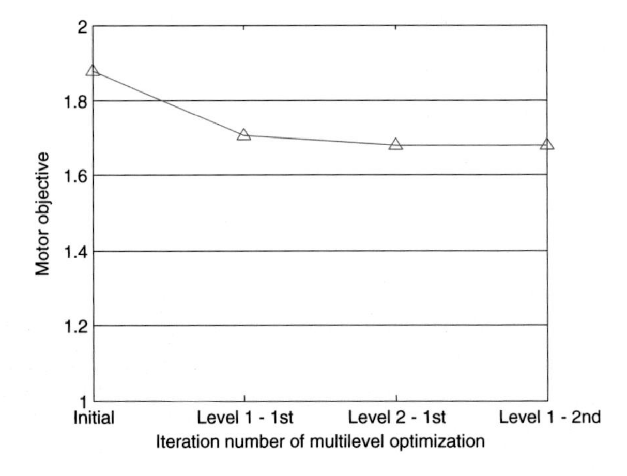

Fig. 6.19 Iteration process of multi-level robust optimization for motor level

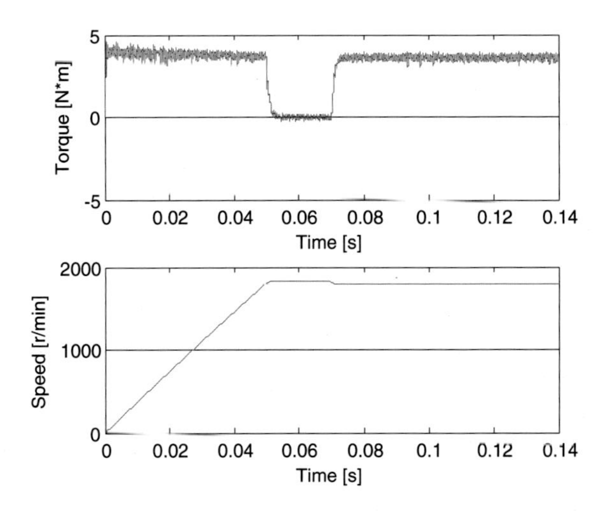

Fig. 6.20 Dynamic performance output by robust optimization

Figure 6.20 illustrates the dynamic performance of drive system by using the algorithm parameters obtained from the robust optimization approach.

After MCA, the reliabilities of all constraints are 1, and the corresponding sigma levels are all more than 6. As a result, the POFs for the motor, controller, and whole system are almost zero. This is much better than those obtained by the deterministic design approach and satisfies the initial reliability requirements. Therefore, the POF of deterministic optimization is bigger than the robust optimization. As a matter of fact, the lower cost of deterministic optimization is obtained at the expense of low reliability and robustness. This is not acceptable in engineering design.

In summary, compared with the initial design, the solutions of robust optimization have many improvements, such as higher output power and efficiency, lower cost, and better reliability and robustness. Meanwhile, the objectives obtained by the deterministic design are smaller than those by the robust design optimization, but the reliabilities, robust levels and POFs of the motor, control, and system levels are obviously worse than the latter. Finally, the obtained control parameters by the robust multi-level optimization are not sensitive to the disturbances of motor output parameters. Therefore, the system dynamic performances can be ensured by the proposed method. This is very valuable for engineering batch production.

6.5 Summary

The manufacturing quality and reliability are two terms used to evaluate the variations of motor performance against the material variations, manufacturing tolerances, and assembling errors, which is very important for the industry. The robust approach based on DFSS has been presented for the design optimization of electrical machines and drive systems so as to improve their manufacturing quality and reliability in mass production in this chapter. Several cases of single- and multi-objective optimization are investigated. The multi-level robust design optimization method has been presented for the system-level design optimization of electrical drive systems. From the investigation of the design example, it can be seen that this approach can significantly improve the drive system reliability, which will benefit the manufacturing of those devices and extent their applications in industry. Therefore, the robust design optimization is a valuable and necessary step for design optimization of electromagnetic devices and systems from the perspective of engineering design, which can improve the product reliability and quality, and save the design cost and cycle.

References

1. Lei G, Wang TS, Guo YG, Zhu JG, Wang SH (2014) System level design optimization method for electrical drive system: deterministic approach. IEEE Trans. Ind. Electron. 61 (12):6591–6602
2. Lei G, Wang TS, Zhu JG, Guo YG, Wang SH (2015) System level design optimization method for electrical drive system: robust approach. IEEE Trans Ind Electron 62(8):4702–4713
3. Lei G, Guo YG, Zhu JG et al (2012) System level six sigma robust optimization of a drive system with PM transverse flux machine. IEEE Trans Magn 48(2):923–926
4. Khan MA, Husain I, Islam MR, Klass JT (2014) Design of experiments to address manufacturing tolerances and process variations influencing cogging torque and back EMF in the mass production of the permanent-magnet synchronous motors. IEEE Trans Ind Appl 50 (1):346–355
5. Coenen I, Giet M, Hameyer K (2011) Manufacturing tolerances: estimation and prediction of cogging torque influenced by magnetization faults. In: Proceedings of 14th european conference on power electronics and applications, pp 1–9, Aug./Sep. 2011
6. Gasparin L, Fiser R (2011) Impact of manufacturing imperfections on cogging torque level in PMSM. In: Proceedings of 2011 IEEE ninth international conference on power electronics and drive systems (PEDS), pp 1055–1060

7. Lei G, Liu CC, Guo YG, Zhu JG (2015) Multi-disciplinary design analysis and optimization of a PM transverse flux machine with soft magnetic composite core. IEEE Trans Magn 51(11), Article 8109704
8. Vese I, Marignetti F, Radulescu MM (2010) Multiphysics approach to numerical modeling of a permanent-magnet tubular linear motor. IEEE Trans Ind Electron 57(1):320–326
9. Kreuawan S, Gillon F, Brochet P (2008) Optimal design of permanent magnet motor using multi-disciplinary design optimisation In: Proceedings of 18th international conference on electrical machines, Vilamoura, pp 1-6, 6−9 Sept 2008
10. Taguchi G, Chowdhury S, Wu Y (2004) Taguchi's quality engineering handbook. Wiley, New Jersey
11. Omekanda AM (2006) Robust torque and torque-per-inertia optimization of a switched reluctance motor using the Taguchi methods, IEEE Trans Ind Appl, 42(2):473−478
12. Kim S-I, Lee J-Y, Kim Y-K et al (2005) Optimization for reduction of torque ripple in interior permanent magnet motor by using the Taguchi method. IEEE Trans Magn 41(5):1796–1799
13. Ashabani M, Mohamed YA, Milimonfared J (2010) Optimum design of tubular permanent-magnet motors for thrust characteristics improvement by combined Taguchi-neural network approach. IEEE Trans Magn 46(12):4092–4100
14. Hwang C-C, Lyu L-Y, Liu C-T, Li P-L (2008) Optimal design of an SPM motor using genetic algorithms and Taguchi method. IEEE Trans Magn 44(11):4325–4328
15. Lee S, Kim K, Cho S et al (2014) Optimal design of interior permanent magnet synchronous motor considering the manufacturing tolerances using Taguchi robust design. IET Electr Power Appl 8(1):23–28
16. Koch PN, Yang RJ, Gu L (2004) Design for six sigma through robust optimization. Struct. Multidiscip. Optim. 26(3–4):235–248
17. Lei G, Zhu JG, Guo YG, Hu JF, Xu W, Shao KR (2013) Robust design optimization of PM-SMC motors for Six Sigma quality manufacturing. IEEE Trans Magn 49(7):3953–3956
18. Lei G, Zhu JG, Guo YG, Shao KR, Xu W (2014) Multi-objective sequential design optimization of PM-SMC motors for Six Sigma quality manufacturing. IEEE Trans Magn 50 (2):7017704
19. Meng XJ, Wang SH, Qiu J, Zhang QH, Zhu JG, Guo YG, Liu DK (2011) Robust multi-level optimization of PMSM using design for six sigma. IEEE Trans Magn 47(10):3248–3251
20. Gasparin L, Cernigoj A, Markic S, Fiser R (2008) Prediction of cogging torque level in PM motors due to assembly tolerances in mass-production. Int J. Comput Math Elect Electron Eng (COMPEL) 27(4):911–918
21. Gasparin L, Cernigoj A, Markic S, Fiser R (2009) Additional cogging torque components in permanent-magnet motors due to manufacturing imperfections. IEEE Trans Magn 45 (3):1210–1213
22. Islam MS, Islam R, Sebastian T, Ozsoylu S, Chandy A (2010) Cogging torque minimization in PM motors using robust design approach. In: Proceedings of IEEE conference on energy conversion congress and exposition (ECCE), pp 3803−3810
23. Islam MS, Mir S, Sebastian T (2004) Issues in reducing the cogging torque of mass-produced permanent-magnet brushless DC motor. IEEE Trans Ind Appl 40(3):813–820
24. Tjahjono B, Ball P, Vitanov VI et al (2010) Six Sigma: a literature review. Int J Lean Six Sigma 1(3):216–233
25. Reosekar RS, Pohekar SD (2014) Six Sigma methodology: a structured review. Int J Lean Six Sigma 5(4):392–422
26. Aboelmaged MG (2010) Six Sigma quality: a structured review and implications for future research. Int J Quality Reliab Manag 27(3):268–317
27. Wang SH, Meng XJ, Guo NN, Li HB, Qiu J, Zhu JG et al (2009) Multi-level optimization for surface mounted PM machine incorporating with FEM. IEEE Trans Magn 45(10):4700–4703
28. Zhu JG, Guo YG, Lin ZW, Li YJ, Huang YK (2011) Development of PM transverse flux motors with soft magnetic composite cores. IEEE Trans Magn 47(10):4376–4383
29. Guo YG, Zhu JG, Watterson PA, Wei Wu (2006) Development of a PM transverse flux motor with soft magnetic composite core. IEEE Trans Energy Convers 21(2):426–434

30. Guo YG, Zhu JG, Dorrell D (2009) Design and analysis of a claw pole PM motor with molded SMC core. IEEE Trans Magn 45(10):4582–4585
31. Deb K, Pratap A, Agarwal S, Meyarivan T (2002) A fast and elitist multi-objective genetic algorithm: NSGA-II. IEEE Trans Evol Comput 6(2):182–197
32. Yao D, Ionel DM (2013) A review of recent developments in electrical machine design optimization methods with a permanent-magnet synchronous motor benchmark study. IEEE Trans Ind Appl 49(3):1268–1275
33. Mendes MHS, Soares GL, Coulomb J-L, Vasconcelos JA (2013) Appraisal of surrogate modeling techniques: a case study of electromagnetic device. IEEE Trans Magn 49(5):1993–1996
34. Wang LD, Lowther DA (2006) Selection of approximation models for electromagnetic device optimization. IEEE Trans Magn 42(2):1227–1230
35. Meng XJ, Wang SH, Qiu J, Zhu JG, Wang Y, Guo YG et al (2010) Dynamic multi-level optimization of machine design and control parameters based on correlation analysis. IEEE Trans Magn 46(8):2779–2782
36. Hasanien HM, Abd-Rabou AS, Sakr SM (2010) Design optimiz-ation of transverse flux linear motor for weight reduction and perfo-rmance improvement using response surface methodology and GAs. IEEE Trans Energy Convers 25(3):598–605
37. Hasanien HM (2011) Particle swarm design optimization of transverse flux linear motor for weight reduction and improvement of thrust force. IEEE Trans Ind Electron 58(9):4048–4056
38. Bolognani S, Bolognani S, Peretti L, Zigliotto M (2009) Design and implementation of model predictive control for electrical motor drives. IEEE Trans Ind Electron 56(6):1925–1936
39. Morel F, Lin-Shi XF, Retif J-M, Allard B, Buttay C (2009) A comparative study of predictive current control schemes for a permanent-magnet synchronous machine drive. IEEE Trans Ind Electron 56(7):2715–2728
40. Wang TS, Zhu JG, Zhang YC (2011) Model predictive torque control for PMSM with duty ratio optimization. In Proceedings of 2011 international conference on electrical machines and systems (ICEMS), 20-23 Aug. 2011, pp. 1−5

Chapter 7
Application-Oriented Design Optimization Methods for Electrical Machines

Abstract From the perspective of engineering applications, the design optimization of electrical machines and drive systems are generally proposed with several specific requirements and constraints, such as the rated torque, the rated speed, the given volume and mass, etc. Therefore, the corresponding design optimization problems are actually oriented by the applications. This chapter aims to develop an application-oriented design optimization method for electrical machines by the deterministic and robust design approaches, respectively. Two kinds of applications are investigated. The first one is about the design optimization of permanent magnet soft magnetic composite machines for compressor drives in refrigerators and air-conditioners. The second one is about the design optimization of flux-switching permanent magnet machines for hybrid electric vehicle drives.

Keywords Application-oriented design optimization method · Motor topologies · Hybrid electric vehicles · Transverse flux machine · Flux-switching permanent magnet machine

7.1 Introduction

With the fast development of CAD/CAE software, new materials, flexible mechanical manufacturing technologies, and advanced optimization and control algorithms, it is possible to design a motor to meet the special requirements of a particular application.

The electric vehicles and hybrid electric vehicles (HEVs) are good examples, which are attracting great attentions and funding from the governments and general public around the world because of the worldwide fossil fuel energy crisis and severe greenhouse gas emissions. To improve the efficiency and drive performance with reduced volume, weight, and cost of novel electrical machines and drive systems to meet the challenging requirements of HEVs, a great amount of research efforts are being directed towards the development of high performance electrical machines and their drive systems [1–4].

G. Lei et al., *Multidisciplinary Design Optimization Methods for Electrical Machines and Drive Systems*, Power Systems,
DOI 10.1007/978-3-662-49271-0_7

Since each vehicle company has its own design conception, the requirements for drive motors of different companies are different. On the other hand, there are many motor types and design schemes for each application. Therefore, for a given application, all possible motor topologies and structural parameters should be investigated to get a globally optimum among different options.

This chapter presents an application-oriented design optimization method for novel electrical machines for domestic appliances and HEVs, respectively. Section 7.2 presents the proposed optimization method for electrical machines by the deterministic approach. Section 7.3 presents a robust approach for the application-oriented design optimization method with a design example of the plug-in HEV (PHEV) drive, followed by the remarks and a summary.

7.2 Application-Oriented Design Optimization Method

7.2.1 Method Description

Figure 7.1 illustrates a framework for the proposed application-oriented design optimization method for electrical machines by the deterministic approach. It mainly includes the following five steps:

Step 1: Define the specifications in terms of specific applications, such as refrigerators and HEVs, including the rated speed, output power, torque and volume.

Step 2: Determine the design options, such as motor types, topologies and materials. Even when the motor type is chosen, there could be various different options, e.g. the flux-switching permanent magnet machine (FSPMM) and variations of different winding configurations and combination of stator/rotor poles [5–7]. Also, different materials can be employed to design the stator cores of permanent magnet (PM) motors, such as silicon sheet steel and soft magnetic composite (SMC). For PMs, the rear-earth and ferrite magnets are two popular options. Other options include the winding type and cooling methods etc. All these are directly related to the output performance and safe operation of the designed machines, such as torque and temperature rise.

Step 3: Establish an initial design for each option. This step includes the development of multi-disciplinary analysis model, determination of initial dimension and performance evaluation model for each motor option. The analysis model mainly includes the finite element model (FEM) for the evaluation of motor performance, such as electromagnetic torque and temperature distribution.

Step 4: Develop a uniform optimization model for all options and optimize each option to acquire its optimal design parameters and performance by using the optimization methods discussed in Chaps. 3 and 4, such as genetic

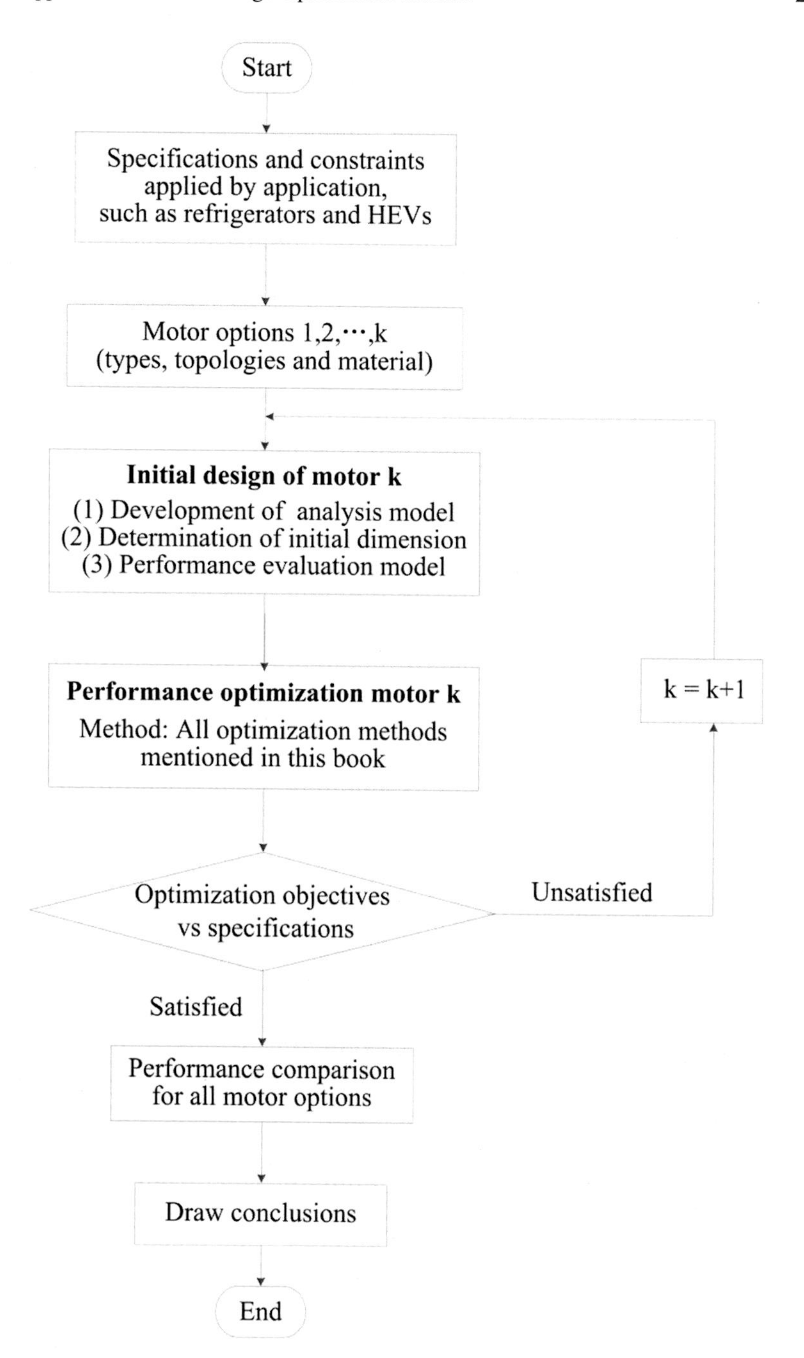

Fig. 7.1 Framework of application-oriented design optimization method for electrical machines

algorithm (GA) and particle swarm optimization (PSO) algorithm [8, 9], and response surface model (RSM) and Kriging model [10–12]. It should be noted that, sequential optimization method can be employed here to improve the optimization efficiency for the low-dimensional design problems, and multi-level optimization method can be employed for high-dimensional design problems [12–17].

Step 5: Compare the optimal results of all options, and output the best one as the final optimal solution for that specific application.

7.2.2 An Optimal PM-SMC Machine for a Refrigerator

Refrigerators are commonly used domestic appliances, and each has different specifications for its drive machine. Several popular ones are the cost, output power and efficiency. Meanwhile, many types of electrical machines have been designed for driving refrigerators so far. Therefore, in order to achieve a best design, all possible motor types and structures should be investigated, and each of them should be optimized for its best performance. The best ones of these motors will be compared to find the final optimal design.

In our previous work, a PM transverse flux machine (TFM) with SMC core as shown in Fig. 2.3 has been developed for the drive machine used in a kind of refrigerator [18, 19]. This refrigerator will be considered as the first application for the proposed application-oriented design optimization method. Table 7.1 lists some of the specifications for this application.

This PM-SMC TFM was designed to deliver an output power of 640 W (or torque of 3.4 Nm) at the rated speed of 1800 rev/min. From the experimental results, it can be seen that this prototype can present good performance [18, 19]. Meanwhile, this motor was compared with two other commercial motors of laminated cores. One is a high efficiency induction motor with the rated torque of 3.72 Nm at 1410 rev/min, 75 % efficiency and 80°C temperature rise in the coil. Its outer diameter is 160 mm and total length is 234 mm. The second one is a radial field NdFeB brushless DC servo motor with rated torque of 3.45 Nm at the rated speed of 3000 rev/min. Its outer diameter is 100 mm and axial length 217 mm. Through the comparison, it can be seen that the proposed SMC motor features a

Table 7.1 Main motor specifications for a refrigerator

Parameter	Unit	Value
Number of phases	–	3
Rated speed	rpm	1800
Rated power	W	640
Rated torque	Nm	3.4
Motor outer diameter	mm	100
Stator axial length	mm	93

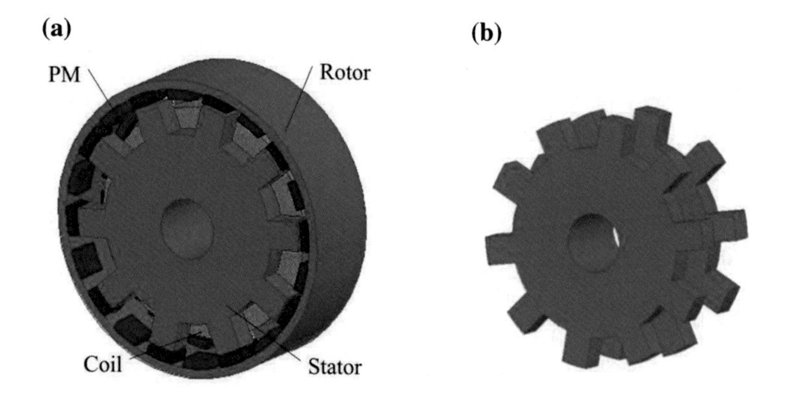

Fig. 7.2 Analysis model in Ansoft, **a** one phase motor stack, and **b** one phase SMC stator stack

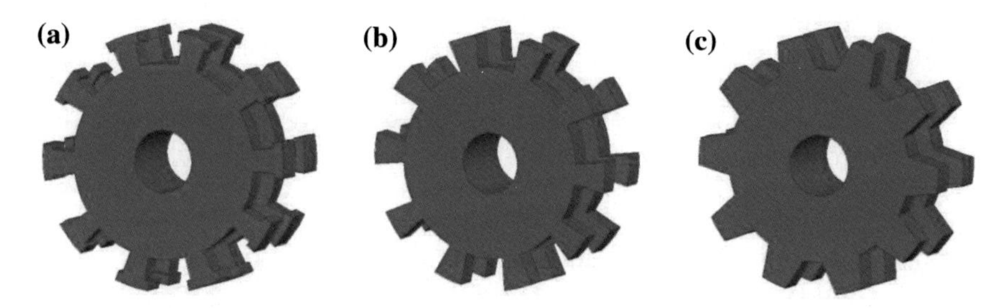

Fig. 7.3 Stator claw pole topologies: **a** rectangular-shoe pole structure, **b** and **c** arc pole structures

torque per unit volume of 4.5 times that of the laminated induction motor. Meanwhile, it delivers 2.25 times the torque per unit volume of the second brushless DC servo motor [18]. Therefore, the proposed PM-SMC TFM has better performances than those two commercial motors.

As shown in Fig. 7.2, only one stator phase stack has been investigated for that PM-SMC TFM in our previous work. In practice, the PM claw pole motor with SMC core may have various other kinds of stator phase stack structures besides the original one shown in Fig. 7.2b. Figure 7.3 shows several possible SMC claw pole stator topologies. The first one has rectangular-shoe poles, which is derived from the design of SMC stator for claw pole motors [20, 21]. The second and third are arc tooth stator. The following work will investigate several stator topologies for this TFM and present an optimal design by using the proposed application-oriented design optimization method.

To obtain the optimal SMC core for this PM TFM, all three topology structures of phase stacks and the dimensions of SMC stator have to be investigated. Meanwhile, in the manufacturing process, besides the density and dimensions of SMC stator, the dimensions of PMs are also important to the machine performance. Therefore, all the factors mentioned above are considered to minimize the material cost and maximize the output power of this TFM. Based on the specifications listed in Table 7.1, the uniform optimization model for these motors can be defined as

$$\begin{aligned} \min: \quad & f(\mathbf{x}) = w_1 \frac{Cost}{C_0} + w_2 \frac{P_0}{P_{out}} \\ \text{s.t.} \quad & g_1(\mathbf{x}) = 0.795 - \eta \le 0 \\ & g_2(\mathbf{x}) = 640 - P_{out} \le 0 \\ & g_3(\mathbf{x}) = sf - 0.8 \le 0 \\ & g_4(\mathbf{x}) = J_c - 6 \le 0 \end{aligned}, \tag{7.1}$$

where w_1 and w_2 are the weighting factors, C_0 and P_0 the material cost and output power of the initial prototype, η, P_{out}, sf and J_c the motor efficiency, output power, fill factor and current density, respectively.

Five parameters are selected as the optimization parameters for these three topology structures of stator phase stacks. Three of them are the circumferential angle, axial width of PMs, and core density. In the case of rectangular-shoe poles, the other two parameters are the axial and circumferential widthes of SMC teeth while the tooth height is fixed as 3 mm. In the case of arc poles, the other two parameters are the inner and outer circumferential angles of SMC teeth. Excluding the claw pole part, the other parameters of the three structures are the same.

Tables 7.2,7.3,7.4,7.5 and 7.6 list the optimization results for this motor in terms of different weighting factors. From the tables, the following conclusions can be drawn:

Table 7.2 Optimization results for w_1 = 0.3 and w_2 = 0.7

Parameter	Unit	Original design	Arc	Rectangular shoe
Cost	$	35.6	40.5	43.6
Pout	W	768	870	878
Efficiency	%	82.2	80.3	81.6
Density	g/cm^3	5.84	5.81	5.80
Objective	–	0.859	0.829	0.849

Table 7.3 Optimization results for w_1 = 0.4 and w_2 = 0.6

Parameter	Unit	Original design	Arc	Rectangular shoe
Cost	$	32.5	35.1	38.4
Pout	W	730	781	802
Efficiency	%	82.2	79.5	80.6
Density	g/cm^3	6.15	6.02	5.91
Objective	–	0.861	0.855	0.876

Table 7.4 Optimization results for $w_1 = 0.5$ and $w_2 = 0.5$

Parameter	Unit	Original design	Arc	Rectangular shoe
Cost	$	28.7	32.6	32.4
Pout	W	672	734	690
Efficiency	%	83.0	79.5	80.0
Density	g/cm^3	6.27	6.13	6.35
Objective	–	0.847	0.858	0.883

Table 7.5 Optimization results for $w_1 = 0.6$ and $w_2 = 0.4$

Parameter	Unit	Original design	Arc	Rectangular shoe
Cost	$	27.1	32.3	30.1
Pout	W	640	726	642
Efficiency	%	83.3	79.5	79.5
Density	g/cm^3	6.33	6.15	6.55
Objective	–	0.789	0.848	0.844

Table 7.6 Optimization results for $w_1 = 0.7$ and $w_2 = 0.3$

Parameter	Unit	Original design	Arc	Rectangular shoe
Cost	$	27.1	32.3	30.2
Pout	W	642	726	645
Efficiency	%	83.3	79.5	79.5
Density	g/cm^3	6.31	6.15	6.54
Objective	–	0.820	0.853	0.865

(1) For the situation that $w_1 = 0.3$ and $w_2 = 0.7$ (Table 7.2), TFM with original design stator has the least material cost of $ 35.6 and the highest efficiency of 82.2 %, TFM with rectangular-shoe stator has the highest output power of 878 W. However, the best topology structure is the TFM with arc-teeth stator. The objective is 0.829, which is the minimal one among them and the corresponding material cost and output power are $ 40.5 and 870 W, respectively.
(2) For the situation that $w_1 = 0.4$ and $w_2 = 0.6$ (Table 7.3), similarly, the best topology structure is still the arc teeth stator. The objective is 0.855, which is the minimal one among them and the corresponding material cost and output power are $ 35.1 and 781 W, respectively.
(3) For the situation that $w_1 = 0.5$ and $w_2 = 0.5$ (Table 7.4), TFM with original design stator has the least material cost of $ 28.7 and the highest efficiency of 83.0 %, TFM with arc teeth stator has the highest output power of 734 W. The best topology structure is the TFM with original design stator. The objective is 0.847.
(4) For the situations that $w_1 = 0.6$ (Table 7.5) and $w_1 = 0.7$ (Table 7.6), similarly, the best topology structure is the TFM with original design stator. The optimal objectives and other motor performances can be seen in those two tables.

Therefore, the TFM with arc-teeth stator is the best one for the first two situations of $w_1 = 0.3$ and 0.4, and the TFM of the original stator is the best one for the latter three situations of w_1 = 0.5, 0.6 and 0.7. The best topology structure and dimension parameters are related to the weighting factors in the optimization model chosen according to the applications. In a situation that the cost is more important than the output power, the TFM with arc teeth stator should be taken as the best design schemes.

7.3 Robust Approach for the Application-Oriented Design Optimization Method

7.3.1 *Method Description*

Figure 7.4 shows the framework for the proposed application-oriented design optimization method based on the robust approach for PM machines. Compared to the framework under deterministic approach, there are two main differences.

The first one is the development of robust analysis model for each option. The robust analysis model mainly includes the determination of variations or noise factors and the manufacturing tolerances or distribution parameters. The robust optimization model can be constructed by using the design for Six-Sigma (DFSS) technique. The second one is that the Monte Carlo analysis (MCA) method is required in this method to evaluate the manufacturing quality of the motors in mass production, e.g. one million products per batch, which mainly includes the mean and standard deviation of variations of motor performance and reliability.

7.3.2 *An Optimal FSPMM for a PHEV Drive*

7.3.2.1 FSPMMs and Topologies

PHEVs have been developed in many countries due to the shortage of fossil fuels. The electric drive system as one of key units in the PHEV plays crucial role for its widely successful commercialization. Figure 7.5 shows a novel PHEV powertrain dependent on one electric machine has been proposed by the University of Technology Sydney (UTS) since 2009 [4, 22]. It consists of an energy storage unit comprising of batteries and super-capacitors, a power control unit including the DC link, DC/DC converters and a back to back inverter/rectifier, an electric machine, functioning as either a motor or a generator (M/G), and an internal combustion engine (ICE) working mainly during fast acceleration to provide the extra torque.

The system operation is governed by a special energy management strategy as illustrated in Fig. 7.6 [22], where SOC stands for the state of charge of the energy storage unit, and EM the electric machine. Initially, it is assumed that the battery and super-capacitor banks are fully charged from the grid, and the capacity of the

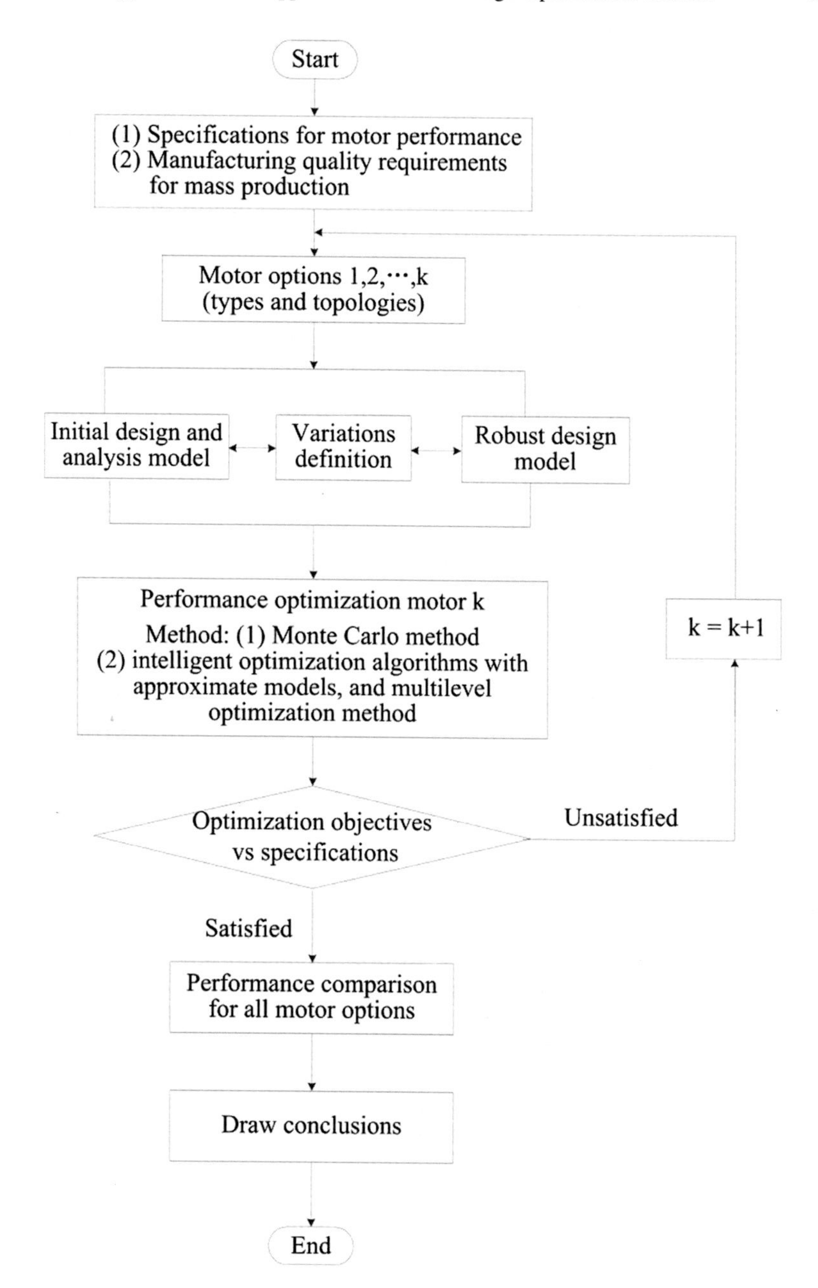

Fig. 7.4 Framework of application-oriented design optimization method based on robust approach

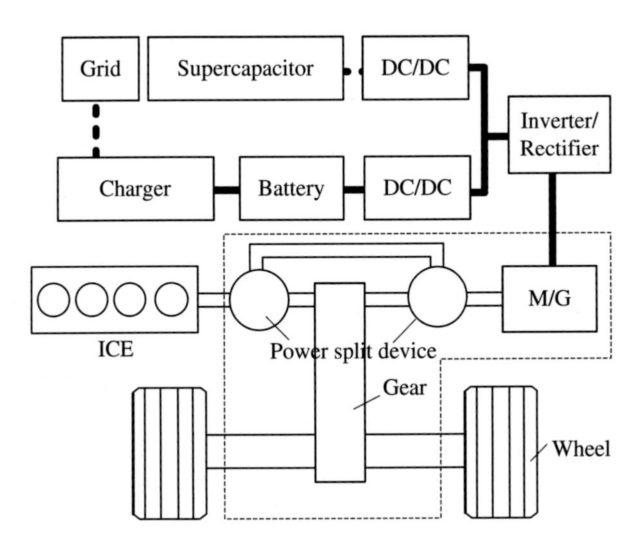

Fig. 7.5 Proposed PHEV configuration

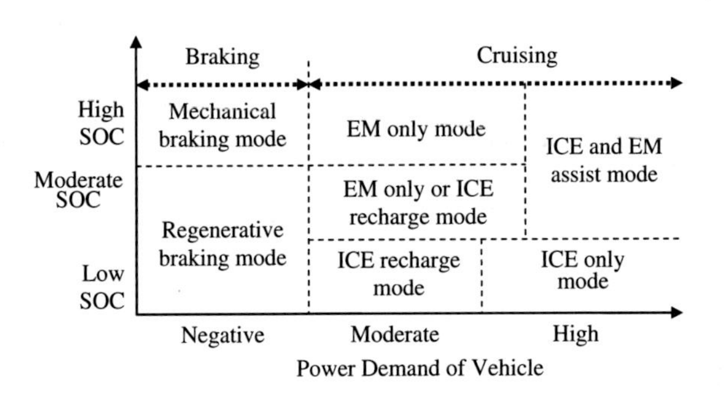

Fig. 7.6 The energy management strategy for proposed PHEV

energy storage is designed such that the car could cover a reasonable long range. In the normal operation mode (high SOC and moderate load), the EM works alone as the prime mover of the car. When it needs extra torque for fast acceleration, the internal combustion engine (ICE) will provide the assistance. When SOC drops, the ICE will recharge the battery through EM while the system is idle. If the load is high and SOC very low, the ICE will work alone to drive the car and recharge battery through EM. It can be seen that the EM in different working conditions has to work continuously. Hence, it must have good attributes of high torque density, high efficiency, strong robustness, and convenience of cooling, etc.

The PHEVs have strict requirements on the drive machine, mainly including high torque/power density, strong flux weakening ability (wide speed range for cruising), good mechanical robustness, strong thermal dissipation capability, etc. [23–26]. Due to these requirements, several FSPMMs with different topology structures have been investigated for a PHEV system in our previous work. Figure 7.7a illustrates the structure of an FSPMM with 12 stator poles and 10 rotor

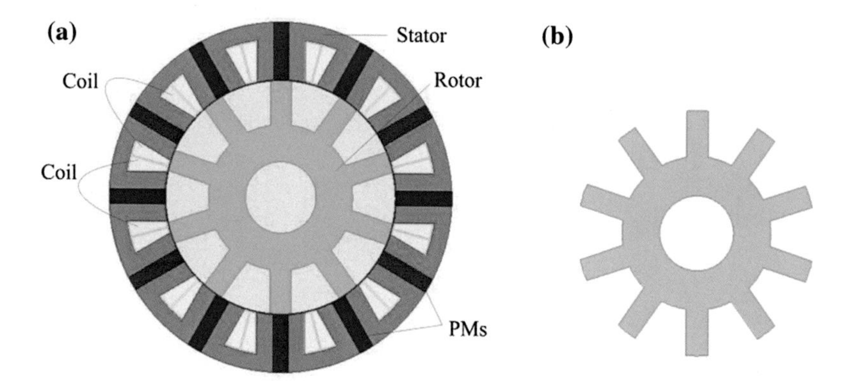

Fig. 7.7 Structures for FSPMM with 12/10 poles, **a** whole motor, and **b** rotor topology

poles (abbreviated as 12/10 FSPMM). Figure 7.7b shows the structure of the rotor. Compared with those traditional PM machines, such as rotor surface mounted or rotor inserted PM machines, FSPMMs have the following main advantages.

(a) Strong thermal dissipation capability—As the PMs are inserted in the stator, they can have greater cross sectional area and are less likely to suffer the demagnetization problem. The winding current density can reach 7–8 A/mm^2 or even larger. In continuous operation, the stator temperature can be maintained well below 125 °C, which is in the range of H-class insulation by water cooling.

(b) Strong structure robustness—Similar to switched reluctance motors, the rotor of FSPMM has no PMs or brushes as shown in Fig. 7.7, and therefore is suitable for high speed operation, e.g. above 20,000 rev/min. For a given power rating, as the rated speed increases, the machine volume can be reduced gradually.

(c) Concentrated winding—The edge connection of stator winding is shorter than distributed ones, which means less copper loss with the same amplitude of stator current.

(d) High power or torque density—Same as the traditional PM machines, PMs in the FSPMM are employed to generate the major air gap flux linkage, and the merit of high power or torque density is retained without extra excited loss [23, 27, 28].

From the extensive research work, it is found that the combination of stator/rotor poles is a very important topology issue for the motor performance [5, 6, 29–31]. Generally, there are many feasible options for the combinations of stator and rotor poles of FSPMMs by the following equation.

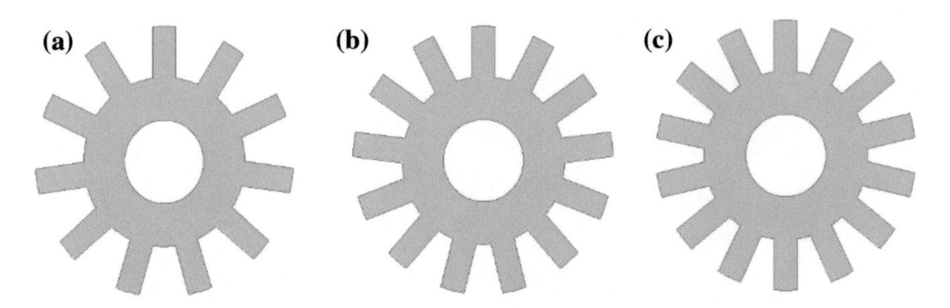

Fig. 7.8 Rotor structures with different poles: **a** 11 poles, **b** 13 poles, and **c** 14 poles

$$\begin{cases} N_s = k_1 m \\ N_r = N_s \pm k_2 \end{cases} \quad (k_1, k_2 = 1, 2, 3\ldots), \tag{7.2}$$

where N_s and N_r are the numbers of stator and rotor poles, respectively, and m is the number of phases. For example, for the FSPMM with 12 stator poles as shown in Fig. 7.7a, there are several promising numbers of rotor poles, such as 10, 11, 13, 14 and 16 poles, respectively, which have been widely investigated in many research works [5, 6, 17, 30]. Figure 7.8 shows three other rotor topology structures for this case. For another example, for the FSPMM with 6 stator poles as shown in Fig. 7.9, several promising numbers of rotor poles are 5, 7 and 8.

On the other hand, there are also some disadvantages for this kind of machine compared with the traditional doubly salient structure machines, such as high cogging torque and odd harmonics in the back electromotive force (EMF). These will reduce the efficiency and increase the torque ripple [7, 22, 27–29]. To overcome these disadvantages, many new topology structures including different combinations of stator/rotor poles, different laminated structures for sheet steels, such as the traditional radially-laminated and new axially-laminated structures [7], and different rotor structures, such as the pole-pairing and pole-skewing [30], have been investigated for the FSPMMs.

Take the FSPMM with 6/7 poles for example. In order to operate at 2000 rev/min, it should be excited by 233 Hz current. As the speed or frequency goes up, the core loss will increase greatly. To reduce this negative influence, a new laminated-structure FSPMM (LSFSPMM) as illustrated in Fig. 7.9b has been proposed in our previous work [7]. Different from the traditional FSPMM, LSFSPMM is laminated axially in parallel to the shaft. As shown, the stator includes 6 respective lamination modules, while the rotor involves 7 modules, which are all made of 0.3 mm high grain oriented silicon steel sheet (HiB). For high magnetic permeability of HiB, each phase flux linkage loops along the lamination and make full use of PMs. Figure 7.10 shows the manufacturing modules for rotor and stator, respectively. Table 7.7 lists the main dimensions for both machines [7].

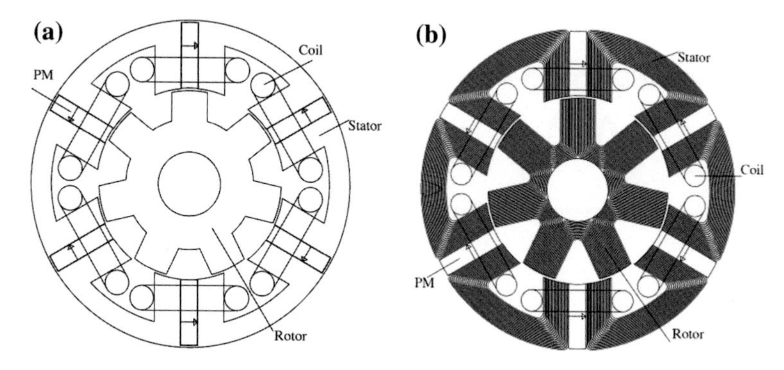

Fig. 7.9 Structure diagrams of two machines with 6/7 poles, **a** traditional FSPMM, and **b** LSFSPMM

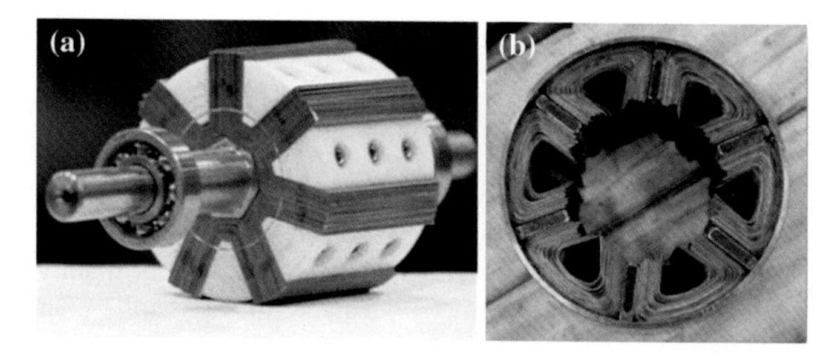

Fig. 7.10 Manufactured modules for the 6/7 LSFSPMM, **a** rotor, and **b** stator

Table 7.7 Main dimensions of two FSPMMs

Items		Unit	Value
Outer radius		mm	49.9
Stator	Yoke height	mm	8.8
	Pole width	mm	21.2
	Pole height	mm	12.4
	Number of turns per pole winding	mm	72
PM	Width	mm	5.4
	Height	mm	19.8
	Relative permeability	mm	1.03
	Magnetic remanence	Tesla	1.19
Air gap length		mm	0.6
Rotor	Pole width	mm	11
	Pole height	mm	13.1
	Yoke height	mm	5.5
Effective axial length		mm	44

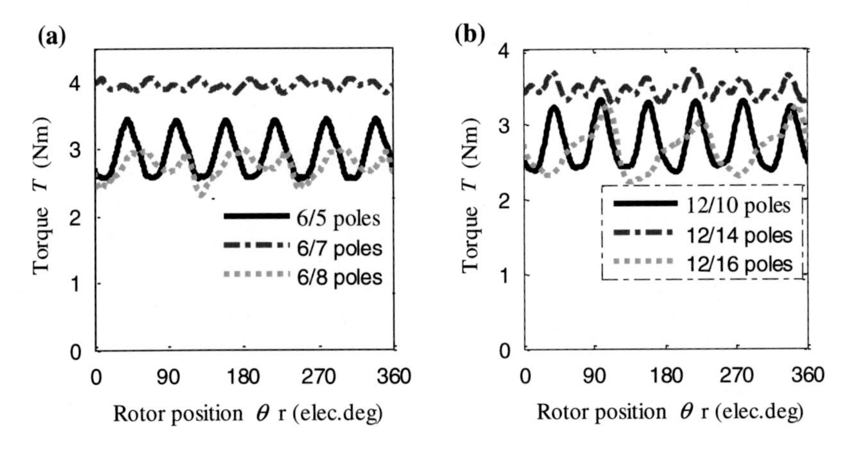

Fig. 7.11 Torque curves versus different combinations of stator/rotor poles, **a** 6-pole stators, and **b** 12-pole stators

Figure 7.11 shows performance comparisons for radially-laminated FSPMMs with different topologies. As shown, 6/7 is the best stator/rotor combination for FSPMM with 6 stator poles, and 12/14 is the best one for FSPMM with 12 stator poles [7].

It should be noted that the above comparison was carried out mainly by experience and with similar structure parameters. This is not fair as the comparison is not based on the optimal design of each combination. Consequently, an interesting problem arising here is that how to accurately seek for an optimal FSPMM among several topology structures for a specific application, e.g. for the drive machine of a specific PHEV/HEV with the given number of stator poles and volume. To solve this problem, the size equation method and some introductory optimization works based on one or two design parameters have been employed [29]. By using these methods, a brief comparison for the performances of different FSPMMs can be obtained.

However, as aforementioned, it is difficult to fairly compare the performances of FSPMMs with different topology structures if each option of them is not fully optimized based on all major structural parameters. The following two sub-sections present a brief qualitative and a quantitative accurate comparison method to solve the above problems through an example of designing an optimal 75 kW FSPMM with 12 stator poles for the drive machine of a PHEV investigated in our previous work [23]. More design requirements for drive machine can be found in that work. As an illustration of the proposed method, two rotor topology structures and motor design parameters are investigated, and their performances are presented and compared.

7.3.2.2 Qualitative Analysis Based on Size Equation

Size equation is a good method to present a qualitative analysis and comparison for FSPMMs with different rotor topology structures. The effect of the combinations of stator/rotor poles on electromagnetic torque can be expressed by the sizing equation as,

$$T_{em} = \frac{\sqrt{2}\pi}{4}\frac{N_r}{N_s}k_s k_d A_s B_g D_{si}^2 l_a c_s \eta \tag{7.3}$$

where k_s and k_d are the skew and leakage factors, respectively, A_s is the armature current electrical loading, B_g the peak flux density at no-load situation, D_{si} the inner diameter of stator, l_a the active stack length, c_s the stator tooth arc factor, and η the efficiency of the machine [32].

As shown, the electromagnetic torque is directly proportional to the number of rotor poles and inversely proportional the number of stator poles. The combinations of stator/rotor poles have also great impacts on the cogging torque. As stated in [32], the period number and magnitude of cogging torque in the FSPMMs can be briefly evaluated by the least common multiple (LCM) and greatest common divisor (GCD) of N_r and N_s, respectively. A higher LCM and a lower GCD have been suggested for effective decrease of the cogging torque.

Table 7.8 tabulates the comparison of FSPMMs with two different combinations of stator/rotor poles based on the size equation. It can be seen that the FSPMM with 12/14 structure has the higher LCM and ratio of N_r over N_s, which can be regarded as the index of smaller cogging torque and bigger magnitude of electromagnetic torque.

As aforementioned, Table 7.8 only shows a brief comparison for FSPMMs with different rotor topology structures. Since the efficiency and structural parameters are involved in (7.3) but not included in the performance evaluation in this table, an accurate quantitative analysis method involving both topology structures and structural parameters are required. This is the main aim of the next sub-section.

7.3.2.3 Quantitative Analysis Based on Optimization

The proposed quantitative analysis is based on the proposed application-oriented design optimization method in Sect. 7.3. It includes three steps, construction of uniform optimization model and its robust form for all topology structures of FSPMMs, development of optimization methods for all design parameters, and quantitative comparison for all performance parameters. The comparison is based on the optimized

Table 7.8 Qualitative analysis results for FSPMMs with 12 stator poles

N_s	N_r	N_r/N_s	*LCM*	*GCD*
12	10	0.83	60	2
12	14	1.17	84	2

performances for all topology structure of FSPMMs for the same objectives and constraints. Therefore, the final optimal FSPMM will possess not only the best topology but also the optimal structural parameters for this topology structure.

A. Uniform Optimization Model

As an example, a FSPMM with two topology structures will be investigated to illustrate the efficiency of the proposed method. The FSPMM is designed to deliver 75 kW output power at a rated speed of 3000 rev/min for a drive machine proposed in the UTS PHEV [23]. For the optimization, objectives are minimizing cogging torque (T_{cog}) and torque ripple (T_{rip}) and maximizing average torque (T_{ave}); and objective function has the form as

$$\min : f(x) = \frac{T_{cog}}{T_{cog\_initial}} + \frac{T_{ave}}{T_{ave\_initial}} + \frac{T_{rip}}{T_{rip\_initial}}$$
$$s.t. \begin{cases} g_1(\mathbf{x}) = 0.9 - \eta \le 0 \\ g_2(\mathbf{x}) = sf - 0.6 \le 0 \\ g_3(\mathbf{x}) = 220 - T_{ave} \le 0 \\ g_4(\mathbf{x}) = T_{rip} - 25 \le 0 \end{cases}, \tag{7.4}$$

where the subscript 'initial' means the values calculated from the initial design scheme as shown in Table 7.9, η is the efficiency, and sf the slot filling factor.

Considering the manufacturing tolerances, the robust optimization model of (7.4) can be obtained within the framework of design for Six-Sigma (DFSS) shown in the last chapter, and it has the form as

Table 7.9 Initial dimensions for FSPMMs with 12 stator poles

Parameters		Unit	Value
Stator	Yoke height	mm	17.5
	Pole width	mm	23.4
	Pole height	mm	16.8
	Turns of winding	turns	6
	Slot filling factor	%	60
	Current density	A/mm^2	7
PM	Width	mm	8
	Height	mm	33.8
	Relative permeability	–	1.03
	Magnetic remanence	T	1.19
Air gap length		mm	0.6
Rotor	Pole width	mm	12
	Pole height	mm	24.1
	Yoke height	mm	17.5

$$\begin{aligned} \min: \quad & \mu_f(\mathbf{x}) \\ \text{s.t.} \quad & \mu_{g_i}(\mathbf{x}) + n\sigma_{g_i}(\mathbf{x}) \leq 0, \; i = 1, \ldots, 4 \end{aligned} \tag{7.5}$$

where μ and σ are the mean and standard deviation of the corresponding terms estimated by MCA, n is the sigma level and is selected as 6 in this work to guarantee that the obtained optimal design can achieve the Six-Sigma quality in industry manufacturing, namely 3.4 defects per million products. To obtain the mean and standard deviation in this equation, the sample size in the MCA is defined as 10,000 [33–36].

From previous experience, it is found that four parameters including depth of stator pole, width of rotor pole, height of rotor yoke and width of permanent magnet are significant to the motor performance. Therefore, they are selected as the optimization parameters in the optimization. For the robust optimization, 0.05 mm was chosen to be the manufacturing tolerances for all these parameters, which means that the standard deviations are 0.05/3 mm. To calculate the reliability of different motors, a term known as probability of failure (POF) will be calculated as follows

$$\text{POF} = \prod_{i=1}^{4} \text{Prob}(g_i \leq 0) \tag{7.6}$$

This POF will be used to calculate the defect rate in the 10,000 products used in the MCA.

B. Optimization Results and Comparison

Figure 7.12 illustrates the comparison of cogging torque curves of 12/10 and 12/14 FSPMMs after robust optimization of (7.5). As shown, the minimal and maximal cogging torques for 12/10 FSPMM are –9.27 and 8.95 Nm, respectively, while they are −6.95 and 8.52 Nm, respectively for 12/14 FSPMM. Therefore, 12/14 has smaller amplitudes of cogging torques.

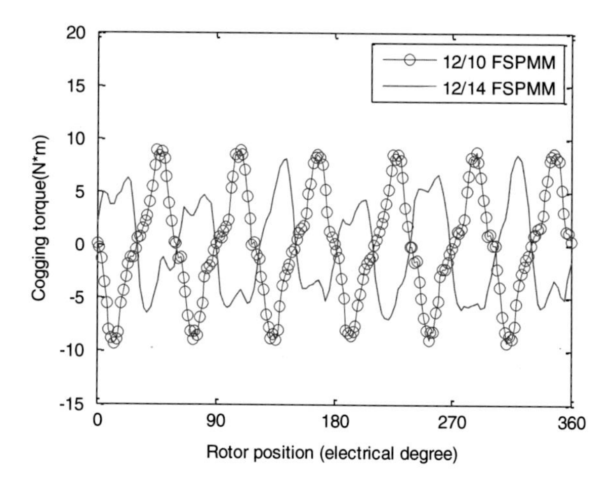

Fig. 7.12 Cogging torque curves after robust optimization

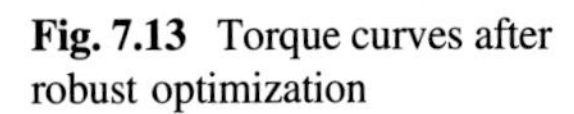

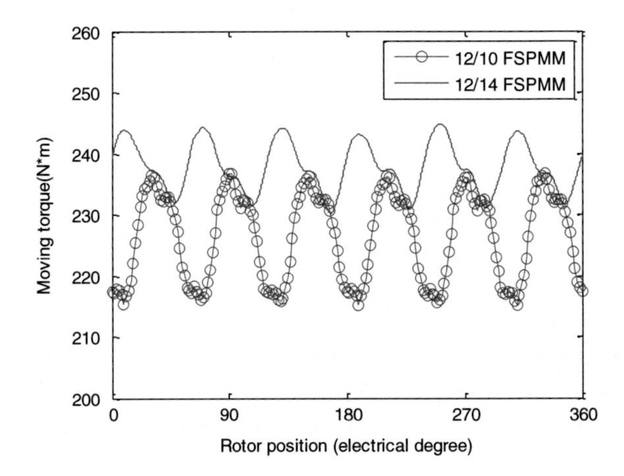

Fig. 7.13 Torque curves after robust optimization

Figure 7.13 shows the comparison of torque curves of 12/10 and 12/14 FSPMMs at rated speed after robust optimization. As shown, the minimal and maximal torques for the 12/10 FSPMM are 215.09 and 236.71 Nm, respectively, and therefore the average torque is 225.90 Nm, the torque ripple is 21.62 Nm (or 9.57 %). The minimal and maximal torques for the 12/14 FSPMM are 230.18 and 244.89 Nm, respectively. Thus, the average torque is 237.54 Nm, and the torque ripple is 14.71 Nm (or 6.19 %). It can be seen that the average torque of the optimal 12/14 motor is higher than that of the optimal 12/10 motor, while the torque ripple is smaller than that of 12/10 motor.

Meanwhile, the reliability and POF of both optimal 12/10 and 12/14 FSPMMs can be calculated by using (7.6) and MCA. It can be found both motors' reliabilities are almost 100 % after calculation.

In conclusion, 12/14 FSPMM has better performance parameters compared with those of 12/10 motor in terms of average torque, torque ripple and cogging torque. It should be noted that only two rotor topologies were investigated in that example. However, other topology structures can be easily investigated and compared by using this method, including different stator/rotor pole combinations and different laminated structures of the steel sheets.

7.4 Summary and Remarks

This chapter presents an application-oriented design optimization method for PM motors by the deterministic and robust approaches, respectively. Two practical applications are investigated as well, namely the drive machines for refrigerators and the UTS PHEV. In the proposed method, both motor topology structures and dimension parameters are investigated to acquire the best performance for each design option.

On the other hand, several remarks could be presented here for this application-oriented design optimization method. The first also the most important one is that only one discipline (electromagnetic analysis) has been investigated for these examples. However, system-level and integrated design optimizations should be the starting points of this application-oriented design optimization method, which means that besides the performance of motor itself, the integrated performance of the whole drive system as well as the whole appliance should be investigated. Therefore, multi-disciplinary design analysis is more important from the perspective of industry applications, particularly the integration of motor and control systems, and should be involved for the application-oriented design optimization method.

The second one is that only five structural parameters (TFM example) and four parameters (FSPMM example) are investigated in this work. For the high-dimensional optimization situation, the multi-level (robust) optimization method presented in Chaps. 4–6 should be employed to improve the optimization efficiency. Due to the high efficiency of the multi-level optimization method, it is able to efficiently optimize and accurately compare the optimal performances among several electrical machines and drive systems with different topologies and parameters.

The third one is that only single objective has been considered in these examples. For the multi-objective optimization situation, the multi-objective sequential optimization strategy presented in Chap. 4 can be taken to improve the optimization efficiency. Therefore, all the developed design optimization methods as shown in Chaps. 4–6 can be employed for the proposed application-oriented design optimization method.

References

1. Zhu ZQ, Howe D (2007) Electrical machines and drives for electric, hybrid, and fuel cell vehicles. Proc IEEE 95(4):746–765
2. Emadi A, Lee YJ, Rajashekara K (2008) Power electronics and motor drives in electric, hybrid electric, and plug-in hybrid electric vehicles. IEEE Trans Ind Electron 55(6):2237–2245
3. Lei G, Wang TS, Guo YG, Zhu JG, Wang SH (2014) System level design optimization method for electrical drive system: deterministic approach. IEEE Trans Ind Electron 61 (12):6591–6602
4. Salisa AR, Zhang N, Zhu JG (2011) A comparative analysis of fuel economy and emissions between a conventional HEV and the UTS PHEV. IEEE Trans Veh Techno 60(1):44–54
5. Chen JT, Zhu ZQ (2010) Winding configurations and optimal stator and rotor pole combination of flux-switching PM brushless AC machines. IEEE Trans Energy Convers 25 (2):293–302
6. Chen JT, Zhu ZQ, Iwasaki S, Deodhar RP (2011) Influence of slot opening on optimal stator and rotor pole combination and electromagnetic performance of switched-flux PM brushless AC machines. IEEE Trans Ind Appl 47(4):1681–1691

7. Xu W, Zhu JG, Guo YG et al (2011) New axial laminated-structure flux switching permanent magnet machine with 6/7 poles. IEEE Trans Magn 47(10):2823–2826
8. Hasanien HM, Abd-Rabou AS, Sakr SM (2010) Design optimization of transverse flux linear motor for weight reduction and performance improvement using response surface methodology and genetic algorithms. IEEE Trans Energy Convers 25(3):598–605
9. Hasanien HM (2011) Particle swarm design optimization of transverse flux linear motor for weight reduction and improvement of thrust force. IEEE Trans Ind Electron 58(9):4048–4056
10. Yao D, Ionel DM (2013) A review of recent developments in electrical machine design optimization methods with a permanent magnet synchronous motor benchmark study. IEEE Tran Ind Appl 49(3):1268–1275
11. Lebensztajn L, Marretto CAR, Costa MC, Coulomb J-L (2004) Kriging: a useful tool for electromagnetic device optimization. IEEE Trans Magn 40(2):1196–1199
12. Lei G, Shao KR, Guo YG, Zhu JG, Lavers JD (2008) Sequential optimization method for the design of electromagnetic device. IEEE Trans Magn 44(11):3217–3220
13. Lei G, Shao KR, Guo YG, Zhu JG, Lavers JD (2009) Improved Sequential optimization method for high dimensional electromagnetic optimization problems. IEEE Trans Magn 45 (10):3993–3996
14. Lei G, Yang GY, Shao KR, Guo YG, Zhu JG, Lavers JD (2010) Electromagnetic device design based on RBF models and two new sequential optimization strategies. IEEE Trans Magn 46(8):3181–3184
15. Lei G, Shao KR, Guo YG, Zhu JG (2012) Multi-objective sequential optimization method for the design of industrial electromagnetic devices. IEEE Trans Magn 48(11):4538–4541
16. Lei G, Guo YG, Zhu JG, Chen XM, Xu W (2012) Sequential subspace optimization method for electromagnetic devices design with orthogonal design technique. IEEE Trans Magn 48 (2):479–482
17. Lei G, Xu W, Hu JF, Zhu JG, Guo YG, Shao KR (2014) Multi-level design optimization of a FSPMM drive system by using SSOM. IEEE Trans Magn 50(2), Article no. 7016904
18. Guo YG, Zhu JG, Watterson PA, Wu W (2006) Development of a PM transverse flux motor with soft magnetic composite core. IEEE Trans Energy Conver 21(2):426–434
19. Zhu JG, Guo YG, Lin ZW et al (2011) Development of PM transverse flux motors with soft magnetic composite cores. IEEE Trans Magn 47(10):4376–4383
20. Guo YG, Zhu JG, Dorrell D (2009) Design and analysis of a claw pole PM motor with molded SMC core. IEEE Trans Magn 45(10):4582–4585
21. Huang YK, Zhu JG et al (2009) Thermal analysis of high-speed SMC motor based on thermal network and 3D FEA with rotational core loss included. IEEE Trans Magn 45(106): 4680–4683
22. Xu W, Lei G, Zhu JG, Guo YG (2012) Theoretical research on new laminated structure flux switching permanent magnet machine for novel topologic plug-in HEV. IEEE Trans Magn 48 (11):4050–4053
23. Xu W, Zhu JG, Zhang YC, Wang TS (2011) Electromagnetic design and performance evaluation on 75 kW axially laminated flux switching permanent magnet machine. Proc ICEMS, pp 1–6
24. Cao R, Mi C, Cheng M (2012) Quantitative comparison of flux-switching permanent-magnet motors with interior permanent magnet motor for EV, HEV and PHEV applications. IEEE Trans Magn 48(8):2374–2384
25. Dorrell DG, Knight AM, Evans L, Popescu M (2012) Analysis and design techniques applied to hybrid vehicle drive machines—assessment of alternative IPM and induction motor topologies. IEEE Trans Ind Electron 59(10):3690–3699
26. Takeno M, Chiba A, Hoshi N, Ogasawara S, Takemoto M, Rahman MA (2012) Test results and torque improvement of the 50-kW switched reluctance motor designed for hybrid electric vehicles. IEEE Trans Ind Appl 48(4):1327–1334
27. Zhu ZQ, Chen JT (2010) Advanced flux-switching permanent magnet brushless machines. IEEE Trans Magn 46(6):1447–1453

28. Hua W, Cheng M, Zhu ZQ, Howe D (2008) Analysis and optimization of back EMF waveform of a flux-switching permanent magnet motor. IEEE Trans Energy Conver 23 (3):727–733
29. Fei W, Luk PCK, Shen JX, Wang Y, Jin M (2012) A novel permanent-magnet flux switching machine with an outer-rotor configuration for in-wheel light traction applications. IEEE Trans Ind Appl 48(5):1496–1506
30. Xu W, Zhu JG, Zhang YC, Hu JF (2011) Cogging torque reduction for radially laminated flux-switching permanent magnet machine with 12/14 poles. In: Proceedings of the 37th Annual Conference on IEEE Industrial Electronics Society (IECON), pp 3590–3595
31. Xu W, Lei G, Zhang Y, Zhu JG et al (2011) Development of electrical drive system for the UTS PHEV. In: Proceedings of the IEEE Energy Conversion Congress and Exposition (ECCE), pp 1886–1893
32. Hua W, Cheng M, Zhu ZQ, Howe D (2006) Design of flux-switching permanent magnet machine considering the limitation of inverter and flux-weakening capability. In: Proceedings of the 41st IAS Annual Meeting—Industry Applications Conference, vol. 5, pp 2403–2410
33. Lei G, Guo YG, Zhu JG et al (2012) System level six sigma robust optimisation of a drive system with PM transverse flux machine. IEEE Trans Magn 48(2):923–926
34. Lei G, Zhu JG, Guo YG, Hu JF, Xu W, Shao KR (2013) Robust design optimization of PM-SMC motors for Six Sigma quality manufacturing. IEEE Trans Magn 49(7):3953–3956
35. Lei G, Zhu JG, Guo YG, Shao KR, Xu W (2014) Multi-objective sequential design optimization of PM-SMC motors for six sigma quality manufacturing. IEEE Trans Magn 50(2):701−720
36. Lei G, Wang TS, Guo YG, Zhu JG, Wang SH (2015) System level design optimization method for electrical drive system: robust approach. IEEE Trans Ind Electron 62(8):4702–4713

Chapter 8
Conclusions and Future Works

Abstract This chapter concludes the book. In summary, this book has focused on the development of new efficient design optimization methods for novel high-performance electrical machines and drive systems under two design approaches, namely the deterministic and robust approaches. These new methods include sequential optimization method and its multi-objective form, multi-level optimization method, multi-level Genetic Algorithm, multi-disciplinary design optimization method and application-oriented system-level design optimization method. To illustrate the efficiency of those proposed methods, several classical test functions, a TEAM benchmark problem, and four kinds of motors have been investigated. As shown, the proposed new design optimization methods can achieve better design objectives for electrical motors and drive systems, such as higher output power and lower material cost, with much smaller computational cost than the traditional methods. The proposed robust design optimization approach can yield optimal designs of electrical drive systems for high quality and high reliability mass production. Based on these investigations and outcomes, several directions have been recommended for the future research.

8.1 Conclusions

After a review of the design fundamentals for electrical machines and drive systems, this book presents several novel efficient design optimization methods in terms of different optimization situations including:

(1) Low-dimensional situation: The sequential optimization method (SOM) was presented for this situation. It consists of two processes, the coarse and fine optimization processes. The main aim of the first process is to reduce the initial big design space to a small one by using the space reduction technique. The main aim of the second process is to find the optimal solution by using local sample updating method. From the investigation of a test function, a TEAM Workshop problem (superconducting magnetic energy storage: SMES) and a permanent magnet (PM) claw pole motor with soft magnetic composite

G. Lei et al., *Multidisciplinary Design Optimization Methods for Electrical Machines and Drive Systems*, Power Systems,
DOI 10.1007/978-3-662-49271-0_8

(SMC) core, it can be found that SOM and its improved form can present better optimal solutions while the required computational cost of finite element analysis (FEA) can be reduced by about 90 %.

(2) Multi-objective situation: The multi-objective sequential optimization method (MSOM) was proposed for this situation. It uses the strategy of SOM and improved central composite design technique to reduce the FEA computational cost. From the investigation of a classic test function and a PM-SMC transverse flux machine (TFM), it can be seen that the obtained Pareto fronts are very close to the exact ones.

(3) High-dimensional situation: The multi-level optimization method based on the sequential subspace optimization strategy and multi-level genetic algorithm (MLGA) was proposed for this situation. Two popular techniques, the local sensitivity analysis and the design of experiments (DOE) techniques have been presented for the sensitivity analysis of design parameters so as to establish the framework of multi-level optimization method. From the investigation of a SMES and a surface-mounted permanent magnetic synchronous machine (SPMSM), it can be found that the multi-level optimization method is efficient and the obtained optimal solutions are better than those obtained by the traditional single-level optimization methods.

(4) Multi-disciplinary situation: A multi-disciplinary analysis, design, and optimization framework was presented for PM motors. The multi-disciplinary analysis includes electromagnetic, thermal, modal and manufacturing analyses. From the investigation of a PM-SMC TFM, it is shown that the obtained optimal solutions are better than the initial design and those obtained by non-multi-disciplinary design optimization methods.

(5) Electrical drive systems situation: The system-level design optimization method based on multi-level optimization strategy and multi-level genetic algorithm is proposed for this situation. The design parameters in both motor and control parts have been optimized at the system level rather than the component level to achieve good steady-state and dynamic performances for the whole drive systems, including larger output power and higher efficiency, and lower material cost and dynamic overshoot. Two examples have been investigated to show the efficiency of the proposed method. The first one is a drive system consisting of a PM-SMC TFM and an improved model predictive control (MPC) system. The second example is composed of an SPMSM and a field oriented control (FOC) system. Through the investigation of these two drive systems, it is found that both the steady-state and dynamic performances of the whole drive systems have been greatly improved, and the computational costs required to obtain the solutions have been reduced significantly.

(6) Mass production situation: The main concerns in mass production of electrical machines and drive systems are the product quality and reliability. There are many unavoidable uncertainties or variations in the industrial manufacturing process of electrical machines and drive systems, including mainly the material diversity, manufacturing error, and assembly inaccuracy, which can result in big variations for the reliability and quality of electrical machines and

drive systems in mass production. The robust optimization approach based on the technique of Design for Six-Sigma (DFSS) has been presented for different design optimization situations to achieve optimal designs of high-performance and high-quality electrical machines and drive systems in mass production. Two multi-level optimization strategies are presented to improve the optimization efficiency for high dimensional problems. From the investigation of a PM-SMC machine and a drive system consisting of this machine and the MPC control scheme, it is found that the reliability and quality level of the investigated electrical machines and drive system have been increased greatly by using the proposed robust approach.

(7) Application-oriented situation: This is the ultimate design optimization target of electrical machines as well as drive systems. The system-level and integrated design optimizations are the two main aspects in this application. A concise application-oriented design optimization method has been presented for only electrical machines in this book with two design approaches, the deterministic and robust approaches. Two examples including an optimal PM-SMC motor for a refrigerator and an optimal FSPMM for the UTS plug-in hybrid electric vehicle (HEV) drive have been investigated to show the efficiency and necessity of the proposed method. It is shown that not only the topology structures but also their dimensional parameters of different motor designs have to be optimized before the final comparison and conclusion so as to find an optimal motor for a specific application.

In summary, the proposed new optimization methods are efficient for design optimization of electrical machines and drive systems. It should be noted that all optimization models for electrical machines are verified by comparing the FEA calculations with experimental results. Therefore, the correctness and efficiencies of the proposed methods have been validated and can be employed for extensive engineering applications.

8.2 Future Works

The design optimization of electrical machines and drive systems is a multi-disciplinary, multi-objective, and multi-level problem and an important and challenging issue in both research and industry communities. In order to find efficient ways to solve this kind of problems, further efforts are required for researchers and engineers coming from different disciplines, including mainly material, electrical engineering, mechanical engineering, quality management and control, and applied mathematics. This section intends to draw a picture for the readers about the trends in this research field. The following six aspects are the recommendations from the authors:

(1) Novel motors with new materials and topologies
Firstly, as more and more new materials are employed for design and manufacturing of electrical machines, new material models are required, including the models of core losses, and thermal and manufacturing properties. Secondly, with these new materials and manufacturing methods, novel topologies will be possible for some types of electrical machines.
(2) Efficient control circuits and algorithms
New topologies for the control circuits are required, which can reduce the feedback time and improve the control accuracy. Meanwhile, improved control algorithms based on current ones are very important for this part, including the improved MPC.
(3) System-level design optimization models
Firstly, the optimization models based on the component level should be established based on the new design models mentioned in the above two parts. Secondly, the system-level multi-disciplinary analysis model should be constructed for the whole appliance or system. For example, the models for the drive systems in HEVs should be integrated with the energy-storage model and power-train model. Meanwhile, the system-level reliability and lifetime models should be constructed for evaluation of the reliability of the motors and drive systems. All these models can be employed to extend the usage and enhance the significance and value of the application-oriented design optimization method.
(4) New design optimization methods
Firstly, the efficiency of MSOM for the high-dimensional multi-objective optimization situation has to be improved. The MSOM based on manifold reduction technique is a promising strategy for this issue.
Secondly, the topology optimization is an important topic in electrical engineering, which should be included in the system-level design optimization.
Thirdly, new modeling methods for high dimensional problems are required, e.g. high dimensional Taylor model.
Finally, the multi-level optimization is based on a series framework in this book. It can be used by combining with current parallel optimization framework to improve the multi-disciplinary optimization efficiency. More interactions between motor and control, for example, the feedback from control system to the motor part, should be investigated as well.
(5) Reliability and quality in mass production
Firstly, the qualitative and quantitative analyses (including probability distributions) of the variations/uncertainties in material modelling, manufacturing and assembly process and control process, and their effects on the performance of the whole drive systems, and new equivalent reliability model for drive systems have to be developed for novel electrical machines and drive systems.
Secondly, some robust control strategies should be investigated for drive systems, such as robust MPC, H_{∞} control, and tolerant control, to improve the reliability of drive systems in the operation environment.

(6) Application
While there are many sophisticated new optimization methods, each with its own advantages and disadvantages, a problem we are facing is to decide which one would be the best for a particular engineering problem. Therefore, a selection strategy should be established and a design platform is required to include all these information. The platform will be a powerful tool for designing and testing novel high performance electrical drive systems with new materials, novel topologies, low cost, high efficiency, reliability and robust system performance, and can greatly shorten the design cycles. Meanwhile, the platform will enable designers to focus their attention on the system performance which they have the best expertise, rather than the mathematical algorithms. This will be very valuable for industry mass production.

Printed in the United States
By Bookmasters